Substance Abuse Treatment: Group Therapy Inservice Training

Based on Treatment Improvement Protocol

TIP 41

U.S. DEPARTMENT OF HEALTH AND HUMAN SERVICES
Substance Abuse and Mental Health Services Administration
Center for Substance Abuse Treatment

1 Choke Cherry Road
Rockville, MD 20857

Acknowledgments

This training manual, based on Treatment Improvement Protocol (TIP) 41, *Substance Abuse Treatment: Group Therapy*, was prepared by the Southeast Addiction Technology Transfer Center for the Substance Abuse and Mental Health Services Administration (SAMHSA), U.S. Department of Health and Human Services (HHS). Catherine D. Nugent, LCPC, served as the Government Project Officer.

The manual was produced under the Knowledge Application program (KAP), contract number 270-09-0307, a joint venture of The CDM Group, Inc., and JBS International, Inc., for SAMHSA, HHS. Christina Currier served as the KAP Contracting Officer's Representative.

Electronic Access and Copies of Publication

This publication may be ordered from the SAMHSA Store at http://www.store.samhsa.gov. Or, please call 1-877-SAMHSA-7 (1-877-726-4727) (English and Espa ol).

The document and accompanying PowerPoint slides can be downloaded from the KAP Web site at http://www.kap.samhsa.gov.

Recommended Citation

Substance Abuse and Mental Health Administration. *Substance Abuse Treatment: Group Therapy Inservice Training*. HHS Publication No. (SMA) SMA-11-4664. Rockville, MD: Substance Abuse and Mental Health Services Administration, 2012.

Originating Office

Quality Improvement and Workforce Development Branch, Division of Services Improvement, Center for Substance Abuse Treatment, Substance Abuse and Mental Health Services Administration, 1 Choke Cherry Road, Rockville, MD 20857.

HHS Publication No. (SMA) 11-4664

Printed 2012

Contents

Training and Manual Overview

Training Purpose

This inservice training manual provides counselors and other clinical staff members with scripted modules to use in trainings for Treatment Improvement Protocol (TIP) 41, *Substance Abuse Treatment: Group Therapy*, published by the Substance Abuse and Mental Health Services Administration's (SAMHSA's) Center for Substance Abuse Treatment (CSAT). The seven training modules will assist program staff in understanding and implementing the evidence-based practices described in TIP 41.

A TIP is the end result of careful consideration of relevant research findings and experiences in clinical settings. For each TIP, a panel of expert clinical researchers, clinical providers, and program administrators (the consensus panel) discusses the issues relevant to the specific TIP. The product of the panel represents the combined and collaborative input of the various viewpoints and provides recommendations for specific best-practice guidelines. The panel's work is reviewed by expert field reviewers. Revisions suggested by these reviewers are incorporated into the final document.

TIP 41 and this training manual present an overview of the role and efficacy of group therapy in substance abuse treatment. The goal of both documents is to offer the latest research and clinical findings and to distill them into practical guidelines for group therapy leaders in the field of substance abuse treatment. The documents describe effective types of group therapy and offer a theoretical basis for group therapy in the treatment of substance use disorders. The information will be a useful guide to supervisors and trainers of beginning counselors, as well as to experienced counselors.

Training Design

The training manual is designed as a seven-module continuing education workshop for substance use disorder counselors and other professionals. Senior staff members and clinical supervisors can easily lead the training sessions.

The modules are between 45 minutes and 1 hour and 45 minutes in length. They can be delivered as stand-alone training sessions or as elements within a large training program. Experienced trainers are encouraged to adjust the schedules based on external factors such as participant skill levels, facility amenities, and other factors that affect training delivery. Trainers should read the corresponding TIP chapters to familiarize themselves with the full content of TIP 41 before presenting a training module.

The primary goal of this training package is to provide a quick, easy, and user-friendly way to deliver the content of TIP 41 to substance abuse treatment providers. The manual covers the types of groups used, criteria for placement in a group, group development, stages of treatment, and group leadership issues, such as leadership styles and strategies for therapy. The trainer should be cognizant of participants' needs and adapt the material to meet these needs. Trainer's notes and suggested talking points are provided to allow flexibility.

Instructional Approach

An experienced substance abuse treatment provider should serve as the trainer for these modules, but no training experience is required to use the materials. The success of the training depends on the willingness of the trainer to use the trainer's notes and PowerPoint slides to enhance discussions to ensure that participants grasp the modules' objectives.

The training generally follows the flow of the TIP. The training can be conducted in small- to medium-sized groups (10ñ25 people).

Materials and Equipment

Each module provides trainer's notes and suggested talking points. Thumbnail copies of the PowerPoint slides that reinforce the topics are provided in the left column. The PowerPoint slides, available at http://www.kap.samhsa.gov/products/trainingcurriculums/index.htm, require a personal computer; they can be saved as presentations or, if necessary, printed to make overhead slides. The training room should be set up to accommodate small groups and comfortable viewing of the PowerPoint slides.

TIP 41 is used as a reference throughout the training. The trainer should order enough copies of TIP 41 to distribute one to each participant. Copies can be ordered free of charge from the SAMHSA Store by telephone at 1-877-SAMHSA-7 (1-877-726-4727) or electronically at http://www.store.samhsa.gov. TIPs can also be downloaded from the Knowledge Application Program (KAP) Web site at http://www.kap.samhsa.gov. *A Quick Guide for Clinicians* based on TIP 41 can also be ordered from the SAMHSA Store or downloaded from the KAP Web site.

References in the TIP have been deleted from the training manual. Trainers and participants should refer to TIP 41 for original sources.

Trainers must have the following materials for all modules:

- Computer
- LCD projector for PowerPoint slides
- Newsprint paper, easel, and colorful markers
- Tape for affixing newsprint to the walls

Manual Format

The start of each module presents the module's learning overview, sections, and objectives. A new discussion topic or activity is designated by a section title and the approximate time needed to complete the section. The left column of the module page displays the following icons to assist the trainer:

Passages in Roman typeface are scripted talking points, which are based on text taken directly from TIP 41. This text can be read verbatim or modified by the trainer. Text in italics typeface provides notes to the trainers such as cues on when to begin a new section.

Learning Objectives

After completing this training, participants will be able to:

- Discuss the use of group therapy in substance abuse treatment. (Module 1)
- Explain five group therapy models and three specialized group therapy models used in substance abuse treatment. (Modules 1 and 2)
- Explain the advantages of group therapy. (Module 1)
- Modify group therapy to treat substance abuse. (Module 1)
- Explain the stages of change. (Module 2)
- Match clients with substance abuse treatment groups. (Module 3)
- Assess clients' readiness to participate in group therapy. (Module 3)
- Determine clients' needs for specialized groups. (Module 3)
- Distinguish differences between fixed and revolving membership groups. (Module 4)
- Prepare clients for groups. (Module 4)
- Describe the tasks for each of the three phases of group development. (Module 4)
- Discuss the importance of making clinical adjustments in group therapy. (Module 5)

- Explain the three stages of treatment. (Module 5)
- Describe the conditions of the early, middle, and late stages of treatment. (Module 5)
- Identify leadership characteristics in the early, middle, and late stages of treatment. (Modules 5 and 6)
- Describe concepts and techniques for conducting substance abuse treatment group therapy. (Module 6)
- Identify training opportunities. (Module 7)
- Appreciate the value of clinical supervision. (Module 7)

Module 1: Groups and Substance Abuse Treatment

Module 1 Overview

The goal of Module 1 is to provide participants with an overview of the training and an overview of group therapy in substance abuse treatment. The information in Module 1 covers Chapter 1 of Treatment Improvement Protocol (TIP) 41, Substance Abuse Treatment: Group Therapy. This module takes 1 hour to complete and is divided into three sections:

- *Welcome, Training Objectives, and Ground Rules (10 minutes)*
- *Presentation: Overview of Groups in Substance Abuse Treatment (45 minutes)*
- *Summary (5 minutes)*

Welcome, Training Objectives, and Ground Rules

10 minutes

After participants have taken their seats, the trainer introduces himself or herself and asks participants to introduce themselves by stating their names, what they do, and one skill they hope to gain from the training.

The trainer distributes copies of TIP 41 to participants, asks them to turn to the table of contents (p. iii) of TIP 41, and instructs them to bring TIP 41 to each training session.

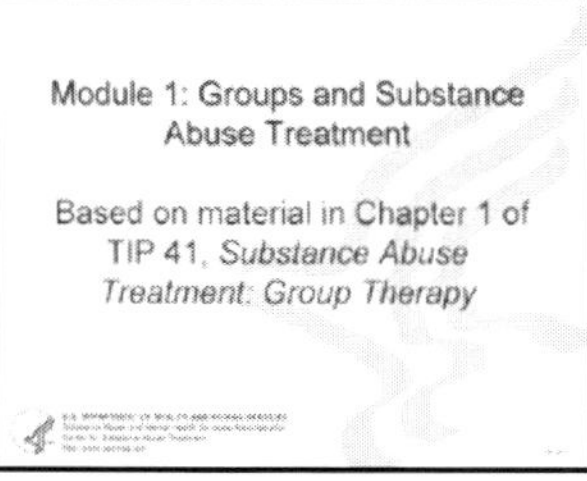

PP 1-1

This training is based on the Substance Abuse and Mental Health Services Administration (SAMHSA)/Center for Substance Abuse Treatment (CSAT) TIP 41, *Substance Abuse Treatment: Group Therapy*. TIP 41 is part of a series of best-practices guidelines developed by SAMHSA/CSAT to assist in providing practical, up-to-date, evidence-based information on important topics in substance abuse treatment. The table of contents for the TIP provides an overview of the training curriculum, which roughly follows the order of the chapters.

Training Objectives
- Discuss the use of group therapy in substance abuse treatment. (Module 1)
- Explain five group therapy models and three specialized group therapy models used in substance abuse treatment. (Modules 1 and 2)
- Explain the advantages of group therapy (Module 1)
- Modify group therapy to treat substance abuse. (Module 1)
- Explain the stages of change. (Module 2)

PP 1-2

The group therapy training is divided into seven modules. Module 1 covers Chapter 1 of TIP 41. After completing all seven modules, you will be able to:

- Discuss the use of group therapy in substance abuse treatment. (Module 1)
- Explain five group therapy models and three specialized group therapy models used in substance abuse treatment. (Modules 1 and 2)
- Explain the advantages of group therapy. (Module 1)
- Modify group therapy to treat substance abuse. (Module 1)
- Explain the stages of change. (Module 2)

Training Objectives (cont.)
- Match clients with substance abuse treatment groups. (Module 3)
- Assess clients' readiness to participate in group therapy. (Module 3)
- Determine clients' needs for specialized groups. (Module 3)
- Distinguish differences between fixed and revolving membership groups. (Module 4)
- Prepare clients for groups. (Module 4)

PP 1-3

- Match clients with substance abuse treatment groups. (Module 3)
- Assess clients' readiness to participate in group therapy. (Module 3)
- Determine clients' needs for specialized groups. (Module 3)
- Distinguish differences between fixed and revolving membership groups. (Module 4)
- Prepare clients for groups. (Module 4)

Training Objectives (cont.)
- Describe the tasks for each of the three phases of group development. (Module 4)
- Discuss the importance of making clinical adjustments in group therapy. (Module 5)
- Explain the three stages of treatment. (Module 5)
- Describe the conditions of the early, middle, and late stages of treatment. (Module 5)
- Identify leadership characteristics in the early, middle, and late stages of treatment. (Modules 5 and 6)

PP 1-4

- Describe the tasks for each of the three phases of group development. (Module 4)
- Discuss the importance of making clinical adjustments in group therapy. (Module 5)
- Explain the three stages of treatment. (Module 5)
- Describe the conditions of the early, middle, and late stages of treatment. (Module 5)
- Identify leadership characteristics in the early, middle, and late stages of treatment. (Modules 5 and 6)

Training Objectives (cont.)
- Describe concepts and techniques for conducting substance abuse treatment group therapy. (Module 6)
- Identify training needs and opportunities (Module 7)
- Appreciate the value of clinical supervision. (Module 7)

PP 1-5

- Describe concepts and techniques for conducting substance abuse treatment group therapy. (Module 6)
- Identify training opportunities. (Module 7)
- Appreciate the value of clinical supervision. (Module 7)

The trainer reviews the ground rules, asks participants whether they would like to add any rules to the list, and writes additional rules on newsprint.

Ground Rules
- Begin and end sessions and breaks on time.
- Respect others and their opinions.
- Allow one person to speak at a time
- Maintain confidentiality.
- Participate in each training session.
- Have fun

PP 1-6

- Begin and end sessions and breaks on time.
- Respect others and their opinions.
- Allow one person to speak at a time.
- Maintain confidentiality.
- Participate in each training session.
- Have fun.

Module 1 Goal and Objectives
Goal
Provide an overview of group therapy used in substance abuse treatment.
Objectives
- Discuss the use of group therapy in substance abuse treatment.
- Define five group therapy models used in substance abuse treatment.
- Explain the advantages of group therapy.
- Modify group therapy to treat substance abuse.

PP 1-7

The goal of Module 1 is to provide an overview of group therapy used in substance abuse treatment. By the end of the session, you will be able to:

- Discuss the use of group therapy in substance abuse treatment.
- Define five group therapy models used in substance abuse treatment.
- Explain the advantages of group therapy.
- Modify group therapy to treat substance abuse.

45 minutes

Presentation: Overview of Group Therapy in Substance Abuse Treatment

The natural propensity of human beings to congregate makes group therapy a powerful therapeutic tool for treating substance abuse one that is as helpful as individual therapy, and sometimes more successful. One reason for this efficacy is that groups intrinsically have many benefits—such as reducing isolation and enabling members to witness the recovery of others and these qualities draw clients into a culture of recovery. Another reason groups work so well is that they are especially suitable for treating problems that commonly accompany substance abuse, such as depression, isolation, and shame.

Although many groups can have therapeutic effects, TIP 41 concentrates only on groups that have trained leaders and that are designed to promote recovery from substance abuse. Emphasis is placed on interpersonal process groups, which help clients resolve problems in relating to other people, problems from which they have attempted to flee by means of addictive substances. This course does not train individuals to be group therapy leaders. Rather, it provides substance abuse counselors with insights and information that can improve their ability to manage the groups they currently lead.

The lives of individuals are shaped by their experiences in groups. People are born into groups; they join groups; they will influence and be influenced by family, social, and cultural groups that constantly shape behavior, self-image, and both physical and mental health.

Group Therapy in Substance Abuse Treatment

- Supports members in times of pain and trouble.
- Enriches members with insight and guidance.
- Is a natural ally with addiction treatment
- Can address factors associated with addiction: depression, anxiety, denial, shame, temporary cognitive impairment, character pathology.
- Has trained leaders.
- Produces healing or recovery from substance abuse.

PP 1-8

Group therapy can support members in times of pain and trouble. A group's therapeutic goals can enrich members with insight and guidance. Group therapy and addiction treatment are natural allies. The effectiveness of group therapy in substance abuse treatment can be attributed to several factors associated with addiction such as depression, anxiety, and denial. Can you name others?

The trainer encourages participants to name other factors and writes them on newsprint.

Most groups in substance abuse treatment have trained leaders. In addition, their goal is to produce healing or recovery from substance abuse.

5 Models of Group Therapy in Substance Abuse Treatment

- Psychoeducational groups: Teach about substance abuse.
- Skills development groups: Hone skills necessary to break free of addiction.
- Cognitive–behavioral groups: Rearrange patterns of thinking/action that lead to addiction.
- Support groups: Provide a forum where members can support constructive change.
- Interpersonal process groups: Enable members to rethink past problems/solutions that led to substance abuse.

PP 1-9

Five group therapy models are frequently used in substance abuse treatment:

- Psychoeducational groups teach about substance abuse.
- Skills development groups help members hone skills necessary to break free of addiction.
- Cognitiveñbehavioral groups encourage members to rearrange patterns of thinking and action that lead to addiction.
- Support groups provide a forum where members can debunk excuses and support constructive change.
- Interpersonal process groups enable members to re-create their past and rethink problems and solutions that led to their substance abuse.

Treatment counselors routinely use the first four models and various combinations of them. The interpersonal process group model is not widely used in substance abuse treatment because of the extensive training required to lead such groups and the long duration of the groups; these groups demand a high degree of commitment from both counselors and clients.

Module 2 provides more details about each type of group.

Treating adult clients in groups has many advantages but can yield poor results if applied indiscriminately or administered by an unskilled or improperly trained leader. There are numerous advantages to using groups in substance abuse treatment. For example:

Advantages of Groups
- Provide positive peer support for abstinence from substances of abuse
- Reduce isolation that people with a substance use disorder experience
- Enable members to witness the recovery of others
- Allow members to see how others deal with similar problems

PP 1-10

- Groups provide positive peer support and pressure for abstinence from substances of abuse. Group therapy elicits commitment by all the group members to attend and to recognize that failure to attend, to be on time, and to treat group time as special disappoints group members and reduces the group's effectiveness.
- Groups reduce the sense of isolation that most people who have substance use disorders experience. They enable participants to identify with others who are struggling with the same issues.
- Groups enable members who abuse substances to witness the recovery of others. From this inspiration, people who are addicted gain hope that they too can maintain abstinence.
- Groups help members learn to cope with their substance use problems by allowing them to see how others deal with similar problems.

Advantages of Groups (cont.)
- Provide information to clients who are new to recovery.
- Provide feedback on group members' values and abilities.
- Offer family-like experiences.
- Encourage, coach, support, and reinforce.

PP 1-11

- Groups can provide useful information to clients who are new to recovery. Members can learn how to avoid triggers for use, the importance of abstinence, and how to self-identify as a person recovering from substance abuse.
- Groups provide feedback on the values and abilities of other group members. This information helps members improve their conception of self or modify faulty, distorted conceptions.
- Groups offer family-like experiences that support and nurture group members. These experiences may have been lacking in the group members' families of origin.
- Groups encourage, coach, support, and reinforce as members undertake difficult or anxiety-provoking tasks.

Modifying Groups
To Treat Substance Abuse
- Specific training and education for group therapy leaders so that they understand therapeutic group work and the special characteristics of clients with substance use disorders
- An understanding of each member's defensive process and character dynamics
- Adaptations to meet the realities of treating clients with substance use disorders

PP 1-12

Some modifications are needed to make group therapy applicable to and effective with clients who abuse substances.

First, group therapy leaders may need specific training and education so that they fully understand therapeutic group work and the special characteristics of clients with substance use disorders. Group therapy is not individual therapy done in a group, nor is it equivalent to 12-Step program practices. Group therapy requires that individuals understand and explore the emotional and interpersonal conflicts that can contribute to substance abuse. The group leader requires specialized knowledge and skill, including a clear understanding of group process and the stages of development of group dynamics.

Second, the individual who is chemically dependent usually comes to treatment with a complex set of defenses and is frequently in denial. The group leader should have a clear understanding of each group member's defensive process and character dynamics.

Third, the theoretical underpinnings and practical applications of general group therapy are not always applicable to individuals who abuse substances. Clients and even staff members often become confused about the different types of group therapy modalities. For instance, the course of treatment may include 12-Step groups, discussion groups, educational groups, continuing care groups, and support groups. Clients can become confused about the purpose of group therapy, and staff can underestimate the impact that group therapy can have. Therefore, the principles of group therapy need to be tailored to meet the realities of treating clients with substance use disorders.

5 minutes

Summary

The trainer:

- *Responds to participants' questions or comments.*
- *Encourages participants to review Chapter 1 of TIP 41.*
- *Instructs participants to read Chapter 2 and reminds them to bring TIP 41 to the next training session.*
- *Reminds participants of the date and time of the next training session.*

Module 2: Types of Groups Used in Substance Abuse Treatment

Module 2 Overview

The goal of Module 2 is to provide participants with an overview of the group therapy models used in substance abuse treatment. The information in Module 2 covers Chapter 2 of Treatment Improvement Protocol (TIP) 41, Substance Abuse Treatment: Group Therapy. *This module takes 1 hour and 45 minutes to complete and is divided into five sections:*

- *Module 2 Goal and Objectives (5 minutes)*
- *Presentation: Stages of Change (10 minutes)*
- *Presentation: Five Group Therapy Models in Substance Abuse Treatment (60 minutes)*
- *Presentation: Three Specialized Group Therapy Models in Substance Abuse Treatment (25 minutes)*
- *Summary (5 minutes)*

5 minutes

Module 2 Goal and Objectives

After participants have taken their seats, the trainer instructs them to turn to Chapter 2 (p. 9) of TIP 41.

Module 2: Types of Groups Used in Substance Abuse Treatment

Based on material in Chapter 2 of TIP 41, *Substance Abuse Treatment: Group Therapy*

PP 2-1

Module 2 covers Chapter 2 of TIP 41.

Module 2 Goal and Objectives
Goal:
Provide details about the group therapy models used in substance abuse treatment.
Objectives:
- Explain the stages of change.
- Describe the five group therapy models used in substance abuse treatment.
- Discuss the three specialized group therapy models used in substance abuse treatment.

PP 2-2

The goal of Module 2 is to provide details about the group therapy models used in substance abuse treatment. The module also explores specialized groups and groups that focus on specific problems. By the end of the session, you will be able to:

- Explain the stages of change.
- Describe the five group therapy models used in substance abuse treatment.
- Discuss the three specialized group therapy models used in substance abuse treatment.

10 minutes

Presentation: Stages of Change

The client's stage of change dictates which group models and methods are appropriate at a particular time.

Stages of Change
- Precontemplation
- Contemplation
- Preparation
- Action
- Maintenance
- Recurrence

PP 2-3

Six stages of change have been identified for individuals with substance use disorders:

- Precontemplation. Individuals are not thinking about changing substance use behaviors and may not consider their substance use a problem.
- Contemplation. Individuals still use substances, but they begin to think about cutting back or quitting substance use.
- Preparation. Individuals still use substances but intend to stop because motivation to quit has increased and the consequences of continued use have become clear. Planning for change begins.
- Action. Individuals choose a strategy for discontinuing substance use and begin to make the changes needed to carry out their plan.
- Maintenance. Individuals work to sustain abstinence and avoid relapse.
- Recurrence. Many will relapse and return to an earlier stage, but they may move quickly through the stages of change and may have gained new insights into problems.

A group comprising members in the action stage who have clearly identified themselves as substance dependent will be far different from the one comprising people who are in the precontemplative stage. Priorities change with time and experience, too. For example, a group of people in their second day of abstinence is very different from a group with 2 years in recovery.

Theoretical orientations also have an impact on the tasks the group is trying to accomplish, what the group leader observes and responds to in group, and the types of interventions that the group leader initiates. Before a group model is used in treatment, the group leader and treatment program should decide on the theoretical framework to be used. Each group model requires different actions from the group leader. Because most treatment programs offer a variety of groups for substance abuse treatment, it is important that these models be consistent with clearly defined theoretical approaches.

60 minutes

Presentation: Five Group Therapy Models in Substance Abuse Treatment

For each type of group, the trainer allows participants to share experiences they have had as members or leaders of that particular group before moving on to the next group.

5 Models of Group Therapy
- Psychoeducational groups
- Skills development groups
- Cognitive–behavioral/problemsolving groups
- Support groups
- Interpersonal process groups

PP 2-4

Substance abuse treatment programs use a variety of group therapy models to meet client needs during the multiphase process of recovery. TIP 41 describes five group therapy models that are effective for substance abuse treatment:

- Psychoeducational groups
- Skills development groups
- Cognitiveñbehavioral/problemsolving groups
- Support groups
- Interpersonal process groups

Each model has something unique to offer to certain populations, and each can provide powerful therapeutic experiences for group members. A model, however, has to be matched with the needs of the particular population being treated; the goals of a particular group are important determinants of the model chosen.

Before beginning the discussion on the types of group models, the trainer asks participants to share their experiences working with different types of groups.

Variable Factors for Groups
- Group or leader focus
- Specificity of the group agenda
- Heterogeneity or homogeneity of group members
- Open-ended or determinate duration of treatment
- Level of leader activity
- Training required for the group leader
- Duration of treatment and length of each session
- Arrangement of room

PP 2-5

Variable factors for the five group models include:

- Group or leader focus
- Specificity of the group agenda
- Heterogeneity and homogeneity of group members
- Open ended or determinate duration of treatment
- Level of leader activity
- Training required for the group leader
- Duration of treatment and length of each session
- Arrangement of room

The trainer instructs participants to turn to Figure 2-2 in TIP 41 (p. 13) and reviews the figure with participants.

We will now take a look at the purpose, principal characteristics, leadership skills and styles, and techniques of each of the five groups.

Psychoeducational Groups: Purpose
- Assist individuals in the precontemplative and contemplative stages of change.
- Help clients in early recovery learn about their disorder.
- Provide family members with an understanding of the behavior of person in recovery.
- Advise clients about other resources and skills that can help in recovery.

PP 2-6

Psychoeducational groups educate clients about substance abuse and related behaviors and consequences. This type of group presents structured, group-specific content, often using videotapes, audiocassettes, and lectures. These groups:

- Assist individuals in the precontemplative and contemplative stages of change. Clients learn to reframe the impact of substance use on their lives, develop an internal need to seek help, and discover avenues for change.
- Help clients in early recovery learn about their disorder. Clients recognize roadblocks to recovery and begin on a path toward recovery.
- Provide family members with an understanding of the behavior of the person in recovery. Families learn how to support their loved one and about their own need for change.
- Advise clients about other resources and skills that can help in recovery. Clients can become familiar with other services such as mutual-help programs and learn skills such as meditation, relaxation, and anger management.

Psychoeducational Groups:
Principal Characteristics
- Work to engage clients in the group discussion and prompt them to relate what they learn to their own substance abuse.
- Are highly structured and often follow a manual or curriculum.

PP 2-7

Psychoeducational groups teach clients that they need to learn to identify, avoid, and eventually master the specific internal states and external circumstances associated with substance use. The coping skills normally taught in skills development groups often accompany this learning. Psychoeducational groups:

- Work to engage participants in the group discussion and prompt them to relate what they learn to their own substance abuse.
- Are highly structured and often follow a manual or curriculum. The leader usually takes a very active role in discussions.

Psychoeducational Groups:
Leadership Skills and Styles
- Understand basic group processes
- Understand interpersonal relationship dynamics
- Have basic teaching skills
- Have basic and some advanced counseling skills

PP 2-8

Leaders in psychoeducational groups primarily assume the roles of educator and facilitator. They also have the same core characteristics of other group leaders: caring, warmth, genuineness, and positive regard for others. Leaders of psychoeducational groups:

- Understand basic group processes   how people interact within a group. They should know how groups form and develop, how group dynamics influence an individual's behavior in group, and how a leader affects group functioning.
- Understand interpersonal relationship dynamics, including how people relate to one another in group settings, how one individual can influence the behavior of others, and how to handle problem group behavior (such as being withdrawn).
- Have basic teaching skills. Such skills include organizing the content, planning for participant involvement in the learning process, and delivering information in a culturally relevant and meaningful way.
- Have basic counseling skills (e.g., active listening, clarifying, supporting, reflecting, attending) and a few advanced counseling skills (e.g., confronting, terminating).

Psychoeducational Groups: Techniques
- Foster an environment that supports participation.
- Encourage participants to take responsibility for their learning.
- Use a variety of learning methods that require sensory experiences.
- Are mindful of cognitive impairments caused by substance use.

PP 2-9

Techniques to conduct psychoeducational groups address how the information is presented and how to assist clients in incorporating learning so that it leads to productive behavior, improved thinking, and emotional change. Techniques:

- Foster an environment that supports participation. Lecturing should be limited, and group discussion should be encouraged.
- Encourage participants to take responsibility for their learning. Leaders should emphasize honest, respectful interactions among all members.
- Use a variety of learning methods that require sensory experiences (e.g., hearing, sight, touch/movement).
- Are mindful of cognitive impairments caused by substance use. People with addictive disorders are known to have subtle, neuropsychological impairments in the early stage of abstinence.

Skills Development Groups: Purpose
- Cultivate skills people need to achieve and maintain abstinence.
- Assume clients lack needed life skills.
- Allow clients to practice skills.
- May either directly relate to substance use or apply to broader areas relevant to recovery.

PP 2-10

Most skills development groups operate from a cognitiveñbehavioral orientation. Many skills development groups incorporate psychoeducational elements into the group process, though skills development may remain the primary goal of the group. These groups:

- Cultivate skills people need to achieve and maintain abstinence.
- Assume clients lack needed life skills.
- Allow clients to practice skills. Clients see how others use the skills and receive positive reinforcement from the group when skills are used effectively.
- May be either directly related to substance use or may apply to broader areas relevant to recovery.

Skills Development Groups: Principal Characteristics
- Have a limited number of sessions and a limited number of participants.
- Strengthen behavioral and cognitive resources.
- Focus on developing an information base on which decisions can be made and actions taken.

PP 2-11

The suitability of a client for a skills development group depends on the unique needs of the individual and the skills being taught in the group. Skills development groups:

- Have a limited number of sessions and a limited number of participants. The group must be small enough to allow members to practice the skills being taught.
- Strengthen behavioral and cognitive resources.
- Focus on developing an information base on which decisions can be made and actions taken.

Skills Development Groups: Leadership Skills and Styles
- Need basic group therapy knowledge and skills.
- Know and can demonstrate skills that clients are trying to develop.
- Are aware of the different ways people approach issues and problems.

PP 2-12

Leaders in skills development groups:

- Need basic group therapy knowledge and skills, such as understanding the ways that groups grow and evolve, knowledge of the patterns that show how people relate to one another in groups, skills in fostering interaction among members, and ability to manage conflicts that arise among members in a group environment.
- Know and can demonstrate skills that clients are trying to develop. Leaders need experience in modeling behavior and helping others learn discrete elements of behavior.
- Are aware of the different ways people approach issues and problems such as anger or assertiveness.

Skills Development Groups: Techniques
- Vary depending on the skills being taught.
- Are sensitive to clients' struggles.
- Hold positive expectations for change and do not shame individuals who seem overwhelmed.
- Depend on the nature of the group, topic, and approach of the group leader.

PP 2-13

The specific techniques used in a skills development group depend on the skills being taught. The process of learning and incorporating new skills may be difficult. Individuals who have been passive and nonassertive may struggle to learn to stand up for themselves. Many changes that seem straightforward have powerful effects at deeper levels of psychological functioning. Techniques:

- Vary depending on the skills being taught.
- Are sensitive to clients' struggles.
- Hold positive expectations for change and do not shame individuals who seem overwhelmed.
- Depend on the nature of the group, topic, and approach of the group leader. Before undertaking leadership of a skills development group, the leader should have participated in the specific skills development group to be led.

Cognitive–Behavioral Groups: Purpose
- Conceptualize dependence as a learned behavior that is subject to modifications through various interventions.
- Work to change learned behavior by changing thinking patterns, beliefs, and perceptions.
- Develop social networks that support abstinence so that the person with dependence becomes aware of behaviors that may lead to relapse and develops strategies to continue in recovery.
- Include psychological elements (e.g., thoughts, beliefs, decisions, opinions, and assumptions).

PP 2-14

Cognitiveñbehavioral groups are a well-established part of the substance abuse treatment field and are particularly appropriate in early recovery. Cognitiveñbehavioral groups use a wide range of formats informed by a variety of theoretical frameworks, but the common thread is cognitive restructuring as the basic methodology of change. Cognitiveñbehavioral groups:

- Conceptualize dependence as a learned behavior that is subject to modifications through various interventions, including identification of conditioned stimuli associated with specific addictive behaviors, avoidance of such stimuli, development of enhanced contingency management strategies, and response desensitization.
- Work to change learned behavior by changing thinking patterns, beliefs, and perceptions.
- Develop social networks that support continued abstinence so that the person with dependence becomes aware of behaviors that may lead to relapse and develops strategies to continue in recovery.
- Include a number of different psychological elements, such as thoughts, beliefs, decisions, opinions, and assumptions. Changing such cognitions and beliefs may lead to greater opportunities to maintain recovery.

Cognitive–Behavioral Groups: Principal Characteristics
- Provide a structured environment within which members can examine the behaviors, thoughts, and beliefs that lead to maladaptive behavior.
- Sometimes follow a treatment manual that provides protocols for intervention techniques.
- Emphasize structure, goal orientation, and a focus on immediate problems.
- Use educational devices.
- Encompass a variety of approaches that focus on changing cognition and the behavior that flows from it.

PP 2-15

Cognitiveñbehavioral groups are often used to address ways a client deals with issues and problems that may be reinforcing substance abuse. These groups:

- Provide a structured environment within which group members can examine the behaviors, thoughts, and beliefs that lead to maladaptive behavior.
- Sometimes follow a treatment manual that provides specific protocols for intervention techniques.
- Emphasize structure, goal orientation, and a focus on immediate problems. Problemsolving groups often have a specific protocol that systematically builds problemsolving skills and resources.
- Use educational devices (e.g., visual aids, role preparation, memory improvement techniques, written summaries, review sessions, homework, audiotapes) to promote rapid and sustained learning.
- Encompass a variety of methodological approaches that focus on changing cognition (beliefs, judgments, and perceptions) and the behavior that flows from it. Some approaches focus on behavior, others on core beliefs, and still others on problemsolving abilities.

Cognitive–Behavioral Groups: Leadership Skills and Styles

- Have a solid grounding in the theory of cognitive–behavioral therapy.
- Are actively engaged in the group and have a consistently directive orientation.
- Allow group members to use the power of the group to develop their own capabilities.
- Recognize, respect, and work with resistance instead of simply confronting it.

PP 2-16

Leaders in cognitiveñbehavioral groups:

- Have a solid grounding in the theory of cognitiveñbehavioral therapy. Training in cognitiveñbehavioral therapy is available. Chapter 7 in TIP 41 provides information on training resources.
- Are actively engaged in the group and have a consistently directive orientation.
- Allow group members to use the power of the group to develop their own capabilities. Leaders may be tempted to become the expert in how to think, how to express that thinking behaviorally, and how to solve problems. It is important not to yield to such temptation.
- Recognize, respect, and work with resistance. Experienced leaders realize that resistance to change is inevitable and can address it without confrontation. TIP 35, *Enhancing Motivation for Change in Substance Abuse Treatment*, has numerous examples of rolling with resistance.

Cognitive–Behavioral Groups: Techniques

- Teach group members about self-destructive behavior and thinking that lead to maladaptive behavior.
- Focus on problemsolving and short- and long-term goal setting.
- Help clients monitor feelings and behavior, particularly those associated with substance use.

PP 2-17

The specific techniques used in cognitive–behavioral groups vary depending on the particular orientation of the leaders. In general, techniques:

- Teach group members about self-destructive behavior and thinking that lead to maladaptive behavior.
- Focus on problemsolving and short- and long-term goal setting.
- Help clients monitor feelings and behavior, particularly those associated with substance use.

Support Groups:
Purpose
- Are useful for apprehensive clients who are looking for a safe environment.
- Bolster members' efforts to develop and strengthen their ability to manage thinking and emotions and to improve interpersonal skills as they recover from substance abuse.
- Address pragmatic concerns.
- Improve members' self-esteem and self-confidence.

PP 2-18

The widespread use of support groups originated in the self-help tradition of the substance abuse treatment field. These groups realize that significant lifestyle change is the long-term goal in treatment and that support groups can play a major role in such life transitions. The focus of support groups can range from strong leader-directed, problem-focused groups in early recovery, which focus on achieving abstinence and managing day-to-day living, to group-directed, emotionally and interpersonally focused groups in later stages of recovery. Support groups:

- Are useful for apprehensive clients who are looking for a safe environment.
- Bolster members' efforts to develop and strengthen their ability to manage thinking and emotions and to improve interpersonal skills as they recover from substance abuse.
- Address pragmatic concerns, such as maintaining abstinence and managing day-to-day living.
- Improve members' self-esteem and self-confidence. The group members and group leader provide specific kinds of support, such as helping members avoid isolation and finding something positive to say about other members' contributions.

Support Groups:
Principal Characteristics
- Often are open ended, with a changing population of members.
- Encourage discussion about members' current situations and recent problems.
- Provide peer feedback and require members to be accountable to one another.

PP 2-19

Support groups always have clearly stated purposes that depend on the members' motivation and stage of recovery. Support groups:

- Often are open ended, with a changing population of members. As new clients move into a particular stage of recovery, they may join a support group appropriate for that stage until they are ready to move on. Groups may continue indefinitely, with new members coming and old members leaving and occasionally returning.
- Encourage discussion about members' current situations and recent problems. Discussion usually focuses on staying abstinent.
- Provide peer feedback and require members to be accountable to one another. In cohesive, highly functioning support groups, member-to member or leader-to-member confrontation can occur.

Support Groups:
Leadership Skills and Style

- Need solid grounding in how groups evolve and how people interact and change in groups.
- Have a theoretical framework that supports group development, members' goals and interactions, and the specific interventions.
- Build connections among members and emphasize what they have in common
- Are usually less directive than for other groups.
- Provide positive reinforcement, model appropriate interactions, respect boundaries, and foster open communication

PP 2-20

Some support groups may be peer generated or led. Leaders are active but not directive. Leaders:

- Need solid grounding in how groups grow and evolve and the ways in which people interact and change in groups.
- Have a theoretical framework in counseling (e.g., cognitiveñbehavioral therapy) that informs their approach to support group development, therapeutic goals for group members, guidance of group members' interactions, and implementation of specific interventions.
- Build connections among members and emphasize what members have in common. It is useful for leaders to have participated in a support group and to have been supervised in support group work before undertaking leadership of such a group.
- Are usually less directive than they are for other types of groups. The leader's primary role is to facilitate group discussion and help group members share their experiences, grapple with their problems, and overcome difficult challenges.
- Provide positive reinforcement, model appropriate interactions, respect individual and group boundaries, and foster open and honest communication.

Support Groups:
Techniques

- Vary with group goals and members' needs.
- Facilitate discussion among members, maintain appropriate group boundaries, help the group work through obstacles and conflicts, and provide acceptance of and regard for members.
- Ensure that interpersonal struggles among group members do not hinder the development of the group or any members.

PP 2-21

Specific group techniques are less important in support groups than they are in other groups, so the leader usually has a less active role in group direction. The goal is to facilitate the evolution of support within the group. Techniques:

- Vary with group goals and members' needs.
- Facilitate discussion among members, maintain appropriate group boundaries, help the group work through obstacles and conflicts, and provide acceptance of and regard for members.
- Ensure that interpersonal struggles among group members do not hinder the development of the group or any members.

Interpersonal Process Groups: Purpose

- Recognize that conflicting forces in the mind, some of which may be outside one's awareness, determine a person's behavior, whether healthful or unhealthful
- Address developmental influences, starting in early childhood, and environmental influences, to which people are particularly vulnerable because of their genetic and other biological characteristics.

PP 2-22

Interpersonal process groups should be led only by well-trained professionals. Today's training provides only an overview. The therapeutic approach of interpersonal process groups focuses on healing by changing basic intrapsychic (within a person) and interpersonal (between people) dynamics. For those people who have become dependent on substances, the interpersonal process group raises and reexamines fundamental developmental issues. As faulty relationship patterns are identified, participants begin to change dysfunctional, destructive patterns. Participants become increasingly able to form mutually satisfying relationships with other people. Interpersonal process groups:

- Recognize that conflicting forces in the mind, some of which may be outside one's awareness, determine a person's behavior, whether healthful or unhealthful.
- Address developmental influences, starting in early childhood, and environmental influences, to which people are particularly vulnerable because of their genetic and other biological characteristics.

Interpersonal Process Groups: Principal Characteristics

- Delve into major developmental issues, searching for patterns that contribute to addiction or interfere with recovery
- Use psychodynamics, or the way people function psychologically, to promote change and healing
- Rely on the here-and-now interactions of members

PP 2-23

Interpersonal process groups:

- Delve into major developmental issues, searching for patterns that contribute to addiction or interfere with recovery. The group becomes a microcosm of the way group members relate to people in their lives.
- Use psychodynamics, or the way people function psychologically, to promote change and healing.
- Rely on here-and-now interactions of members. Of less importance is what happens outside the group or what happened in the past.

Interpersonal Process Groups: Leadership Skills and Styles

- Focus on the present, noticing signs of people re-creating their past in what is going on between and among members of the group
- Monitor how group members relate to one another, how each member is functioning psychologically or emotionally, and how the group is functioning

PP 2-24

Leaders must be trained in psychotherapy. Leaders:

- Focus on the present, noticing signs of people re-creating their past in what is going on between and among members of the group. For example, if a person has a problem with anger, this problem eventually will be reenacted in the group.
- Monitor how group members relate to one another, how each member is functioning psychological or emotionally, and how the group is functioning.

Interpersonal Process Groups: Techniques
- Vary depending on the type of process group and the developmental stage of the group.
- Are based on the needs of group members and the needs of the group as a whole.
- Require a high degree of understanding about and insight into group dynamics and individual behavior.

PP 2-25

In practice, group leaders may use different models at various times and may focus on more than one aspect at a time. For example, a group that focuses on changing the individual will also have an impact on the group's interpersonal relationships and the group as a whole. Techniques:

- Vary depending on the type of process group and the developmental stage of the group.
- Are based on the needs of group members and the needs of the group as a whole.
- Require a high degree of understanding about and insight into group dynamics and individual behavior.

25 minutes

Presentation: Three Specialized Group Therapy Models in Substance Abuse Treatment

For each type of group, the trainer allows participants to share experiences they have had as members or leaders of that particular group before moving on to the next group.

3 Specialized Groups
- Relapse prevention groups
- Communal and culturally specific groups
- Expressive groups (art therapy, dance, psychodrama)

PP 2-26

Three specialized groups, which do not fit into the five model categories, function as unique entities in the substance abuse treatment field:

- Relapse prevention groups
- Communal and culturally specific groups
- Expressive groups (art therapy, dance, psychodrama)

Relapse Prevention Groups: Purpose and Characteristics
- Help clients maintain their recovery by providing them with skills to identify and manage high-risk situations.
- Upgrade the clients' abilities to manage risky situations and stabilize clients' lifestyles through changes in behavior.
- Focus on activities, problemsolving, and skills building.
- Increase clients' feelings of self-control.
- Explore the problems of daily life and recovery.

PP 2-27

Relapse prevention groups focus on helping a client maintain abstinence or recover from relapse. Clients need to achieve a period of abstinence before joining a relapse prevention group. This kind of group is appropriate for clients who are abstinent but cannot necessarily maintain a drug-free state. Relapse prevention groups:

- Help clients maintain their recovery by providing them with skills to identify and manage high-risk situations.
- Upgrade the clients' abilities to manage risky situations and stabilize clients' lifestyles through changes in behavior.
- Focus on activities, problemsolving, and skills building.
- Increase clients' feelings of self-control.
- Explore the problems of daily life and recovery.

Relapse Prevention Groups: Leaders and Techniques
- Monitor client participation for risk of relapse, signs of stress, and need for a particular intervention
- Know how to handle relapse and help the group work through such an event in a nonjudgmental, nonpunitive way
- Understand the range of consequences clients face because of relapse
- Draw on techniques used in cognitive–behavioral, psychoeducational, skills development, and process-oriented groups

PP 2-28

Leaders of relapse prevention groups need a set of skills that are similar to those needed for the skills development group, as well as experience working in relapse prevention. Group leaders:

- Monitor client participation for risk of relapse, signs of stress, and need for a particular intervention.
- Know how to handle relapse and help the group work through such an event in a nonjudgmental, nonpunitive way.
- Understand the range of consequences clients face because of relapse.
- Draw on techniques used in cognitiveñbehavioral, psychoeducational, skills development, and process-oriented groups.

Communal and Culturally Specific Groups
- Build personal relationships with clients before turning to treatment tasks
- Can be integrated into a therapeutic group
- Show respect for a culture and its healing practices

PP 2-29

Communal and culturally specific groups use a specific culture's healing practices and adjust therapy to cultural values. These groups:

- Build personal relationships with clients before turning to treatment tasks.
- Can be integrated into a therapeutic group.
- Show respect for a culture and its healing practices.

Leaders
- Strive to be culturally competent, avoid stereotypes, and allow clients to self-identify
- Are aware of cultural attitudes

PP 2-30

Leaders:

- Strive to be culturally competent, avoid stereotypes, and allow clients to self-identify.
- Are aware of cultural attitudes.

SAMHSA has published several books on topics that help counselors become culturally competent, including TIP 29, *Substance Abuse Treatment for People With Physical and Cognitive Disabilities*; TIP 51, *Substance Abuse Treatment: Addressing the Specific Needs of Women*; and *A Provider's Introduction to Substance Abuse Treatment for Lesbian, Gay, Bisexual, and Transgender Individuals*. These books are available from the SAMHSA Store at http://www.store.samsha.gov. Other resources are listed in Figure 3-7 (p. 48) of TIP 41.

Expressive Groups
- Foster social interaction as group members engage in a creative activity.
- Help clients explore their substance abuse, its origins (e.g., trauma), the effect it has had on their lives, and new options for coping.
- Depend on the form of expression clients are asked to use.

PP 2-31

Expressive groups use therapeutic activities that allow clients to express feelings and thoughts that may be difficult to communicate orally. Expressive groups:

- Foster social interaction as group members engage in a creative activity.
- Help clients explore their substance abuse, its origins (e.g., trauma), the effect it has had on their lives, and new options for coping.
- Depend on the form of expression clients are asked to use.

Expressive Groups: Leaders
- Need to be trained in the specific modality being used (e.g., art therapy, drama therapy).
- Can recognize signs related to histories of trauma and can help clients find the resources they need to work through powerful emotions.
- Are sensitive to a client's ability and willingness to participate in the activity.

PP 2-32

Leaders:

- Need to be trained in the specific modality being used (e.g., art therapy, drama therapy).
- Can recognize signs related to histories of trauma and can help clients find the resources they need to work through powerful emotions.
- Are sensitive to a client's ability and willingness to participate in the activity.

5 minutes

Summary

The trainer:

- *Responds to participants' questions or comments.*
- *Encourages participants to review Chapter 2 of TIP 41.*
- *Instructs participants to read Chapter 3 and reminds them to bring TIP 41 to the next training session.*
- *Reminds participants of the date and time of the next training session.*

Module 3: Criteria for the Placement of Clients in Groups

Module 3 Overview

The goal of Module 3 is to provide participants with an overview of how to match clients with groups depending on clients’ readiness to change and their ethnic and cultural experiences. The information in Module 3 covers Chapter 3 of Treatment Improvement Protocol (TIP) 41, Substance Abuse Treatment: Group Therapy. *This module takes 1 hour to complete and is divided into five sections:*

- *Module 3 Goal and Objectives (5 minutes)*
- *Presentation: Matching Clients With Groups (5 minutes)*
- *Presentation: Assessing Client Readiness for Group (25 minutes)*
- *Presentation: Ethnic and Cultural Experiences in Groups (20 minutes)*
- *Summary (5 minutes)*

5 minutes

Module 3 Goal and Objectives

After participants have taken their seats, the trainer instructs them to turn to Chapter 3 (p. 37) of TIP 41.

Module 3 covers Chapter 3 of TIP 41.

Module 3: Criteria for the Placement of Clients in Groups

Based on material in Chapter 3 of TIP 41, *Substance Abuse Treatment: Group Therapy*

PP 3-1

Module 3 Goal and Objectives

Goal:

Provide an overview of how to match clients with groups, depending on clients' readiness to change and their ethnic and cultural experiences.

Objectives:

- Match clients with substance abuse treatment groups
- Assess clients' readiness to participate in group therapy
- Determine clients' needs for specialized groups

PP 3-2

The goal of Module 3 is to provide an overview of how to match clients with groups, depending on clients' readiness to change and their ethnic and cultural experiences. By the end of the session, you will be able to:

- Match clients with substance abuse treatment groups.
- Assess clients' readiness to participate in group therapy.
- Determine clients' needs for specialized groups.

5 minutes

Matching Clients With Groups

- The client's characteristics, needs, preferences, and stage of recovery
- The program's resources
- The nature of the group or groups available
- The client's ethnic and cultural experiences

PP 3-3

Presentation: Matching Clients With Groups

Matching each individual with the right group is critical for success in group therapy. Before placing a client in a particular group, the counselor should consider:

- The client's characteristics, needs, preferences, and stage of recovery
- The program's resources
- The nature of the group or groups available
- The client's ethnic and cultural experiences

Placement choices are constantly subject to change. Clients may need to move to different groups as they progress through treatment, encounter setbacks, or become more or less committed to recovery. A client can move from a psychoeducational group to a relapse prevention group to an interpersonal process group. The client can also participate in more than one group at the same time.

25 minutes

Assessing Client Readiness for Group

- Begin with a thorough assessment
- Inquire about all drugs used, social networks, and experience with and roles in groups
- Obtain additional information from observation, collateral resources, and other instruments
- Pay attention to relationships at the current stage of recovery
- Recognize when a client is not suited for a group approach or a particular group

PP 3-4

Presentation: Assessing Client Readiness for Group

Placement begins with a thorough assessment of the client's ability to participate in the group and the client's needs and desires for treatment. This assessment begins when the client enters treatment and continues during the initial interview and through as long as the first 4 to 6 weeks of group participation.

The assessment should inquire about all drugs used and look for cross-addictions. The client should be asked about his or her social network and experiences with and roles in groups in the past.

Clinical observation and judgment, information from collateral resources, and findings of assessment instruments should contribute to the decision on a client's readiness and appropriateness for group therapy. Either the group leader or another trained staff person should meet with the client before assignment to a group to evaluate how the client reacts to the group leader and to assess interpersonal relationship experiences. The client can also be observed in a waiting room with other clients to gain insight into how he or she relates to others.

The counselor pays careful attention to a client's relationships at the current stage of recovery because these relationships can reveal the client's ability to participate in groups. Clients need to be able to engage with others.

Not all clients are equally suited for all kinds of groups, nor is any group approach necessary or suitable for all clients with a history of substance abuse. For instance, a person who relapses frequently probably would be inappropriate in a support group of individuals who are in the process of resolving practical life problems. A person who is in the throes of acute withdrawal from crack cocaine does not belong in a group with people who have been abstinent for 3 months. Groups usually can be demographically heterogeneous (e.g., men and women, young and old clients, people of different races and ethnicities), but clients should be placed in groups with people with similar needs.

Clients Who May Be Inappropriate for Group Therapy
- Clients who refuse to participate
- Clients who cannot honor group agreements
- Clients who are unsuitable for group therapy
- Clients in the throes of a life crisis
- Clients who cannot control impulses
- Clients whose defenses would clash with the dynamics of the group
- Clients who experience severe internal discomfort in groups

PP 3-5

Some clients may be inappropriate for group therapy:

- Clients who refuse to participate. No one should be forced to participate in group therapy.
- Clients who cannot honor group agreements. Sometimes, clients can have disqualifying pathologies such as personality disorders or paranoia. In other instances, clients cannot attend groups for logistical reasons such as a conflicting work schedule.
- Clients who are unsuitable for group therapy. Such people might be prone to dropping out or acting in ways contrary to the interests of the group.
- Clients in the throes of a life crisis. Such clients require more concentrated attention than groups can provide.
- Clients who cannot control impulses. Such clients may be suitable for homogeneous groups.
- Clients whose defenses would clash with the dynamics of the group. These people include those who cannot tolerate strong emotions or get along with others.
- Clients who experience severe internal discomfort in groups.

A formal selection process is essential to match clients with the groups best suited to meet their needs. Client evaluators should review completed forms and meet with each candidate for group placement. The evaluator should listen carefully to determine the client's hopes, fears, and preferences.

After specifying the appropriate treatment level, a counselor meets with the client to identify options consistent with this level of care. More specific screens are needed to determine whether the client is appropriate for treatment in a group modality.

Primary Placement Considerations
- Women
- Adolescents
- Level of interpersonal functioning
- Motivation to abstain
- Stage of recovery
- Expectation of success

PP 3-6

TIP 41 identifies several primary placement considerations:

- Women. Studies have shown that women do better in women-only groups than in mixed-gender groups. Women are more likely than men to have experienced traumatic events. Women are less willing to disclose their victimization in mixed-gender groups.

- Adolescents. Local, State, and Federal laws related to confidentiality; infectious disease control; parental permission and notification; child abuse, neglect, and endangerment; and statutory rape are important factors when substance abuse treatment services are delivered to minors. Other complications include school scheduling and the need to include family in the treatment process.

- Level of interpersonal functioning, including impulse control. Two questions to consider when determining a client's level of functioning in a group setting include:

  ñ Does the client pose a threat to others?

  ñ Is the client prepared to engage in the give and take of groups dynamics?

- Motivation to abstain. Clients with low levels of motivation should be placed in psychoeducational groups, which can help them make the transition to the recovery-ready stage.

- Stage of recovery. Different types of groups are appropriate for clients at the different stages of recovery. Figure 3-2 in TIP 41 (p. 43) indicates client placement in specific groups based on the client's stage of recovery. Figure 3-3 in TIP 41 (p. 44) indicates client placement in groups based on the readiness for change model discussed in Module 2.

- Expectation of success. Clients are expected to succeed in the groups. A poor match can be identified early through group monitoring. The group cannot succeed unless each member of the group gets something out of the experience.

20 minutes

Presentation: Ethnic and Cultural Experiences in Groups

Diversity in a Broad Sense
- Defined as differences that distinguish an individual from others and that affect how an individual identifies and how others identify him or her
- Includes age, gender, cultural background, sexual orientation, ability level, social class, education level, spiritual background, parental status, and justice system involvement

PP 3-7

Ethnic and cultural diversity issues take on added importance in a therapeutic group composed of many different kinds of people. As group therapy proceeds, feelings of belonging to an ethnic group can be intensified more than in individual therapy because in the group process the individual may engage many peers who are different, not just the counselor who is different.

Diversity in TIP 41 means differences that distinguish an individual from others and that affect how an individual identifies himself or herself and how others identify him or her. It includes consideration of age, gender, cultural background, sexual orientation, ability level, social class, education level, spiritual background, parental status, and justice system involvement.

Culturally Responsive Group Leaders

- Aware that cultural roles may conflict with treatment requirements.
- Anticipate a particular group's characteristics without automatically assigning them to all individuals in that group.
- Should be open and ready to learn all they can about their clients' cultures.
- Are conscious of how their backgrounds affect their ability to work with a particular population.

PP 3-8

A culturally homogeneous group tends to adopt roles and values from its culture of origin. However, group leaders should be aware that these roles may conflict with treatment requirements. If a group leader believes that cultural traditions might be a factor in a client's participation or in misunderstandings among group members, the leader should check the accuracy of that perception with the client involved. However, individuals cannot always perceive or articulate their cultural assumptions.

Group leaders should anticipate a particular group's characteristics without automatically assigning them to all individuals in that group. For instance, it is a mistake for a program to assign all immigrants to a single group and assume they would be comfortable together.

Leaders should be open and ready to learn all they can about their clients' cultures. Ethnicity and culture have a profound effect on many aspects of treatment.

Group leaders should be conscious of how their own backgrounds affect their ability to work with particular populations. For example, a group therapy leader who has survived domestic violence may have difficulties working with spouse abusers.

The greater the mix of ethnicities in a group, the more likely that biases will emerge and require mediation. A client should not be asked to give up any cultural beliefs, feelings, or attitudes. The client should be encouraged to share these beliefs even though they may upset other members. Although group leaders may be uncomfortable when a member talks about racism and bigotry, such expressions may be an important part of a person's recovery process.

Diversity and Placement

- Address the substance use problem in a manner that is congruent with the client's culture
- Appreciate that particular cultures use substances at specified social occasions
- Assess the behaviors and attitudes of current group members to determine whether a new client would match the group
- Understand personal biases and prejudices about specific cultural groups

PP 3-9

Before placing a client in a particular group, the counselor needs to understand the influence of culture, family structure, language, identity processes, health beliefs and attitudes, political issues, and stigma associated with minority status for each client who is a potential candidate for the group. The counselor needs to:

- Address the substance use problem in a manner that is congruent with the client's culture. For instance, some cultures use substances as part of rituals. This entwinement of substance use and culture does not mean that the counselor cannot discuss the issue of this substance use with a client. Some clients will reduce or eliminate the use of substances once they examine their beliefs and experiences.

- Appreciate that particular cultures use substances, usually in moderation, at specified social occasions. A culturally sensitive discussion of this issue with clients may result in individual decisions to abstain on these occasions, despite cultural pressure to use.

- Assess the behaviors and attitudes of current group members to determine whether a new client would match the group. Because group members are less restricted to their usual social circles and customary ethnic and cultural boundaries, the group is potentially a social microcosm within which members may safely try out new ways of relating. Nevertheless, potential problems between a candidate and existing group members should be identified and counteracted to prevent dropout and promote engagement cohesion among members.

- Understand personal biases and prejudices about specific cultural groups. A group leader should be conscious of personal biases to be aware of countertransference issues, to serve as a role model for the group, and to create group norms that permit discussion of prejudices and other sensitive topics.

Four Processes That Occur Within Multiethnic Groups
- Symbolism and nonverbal communication
- Cultural transference of traits from one person of a certain culture to another person of that culture
- Cultural countertransference, the leader's emotional reactions to a client
- Ethnic prejudice

PP 3-10

Four major processes that occur within multiethnic groups have been identified:

- Symbolism and nonverbal communication. In some cultures, direct expression of thoughts and feelings is considered unseemly; symbolic gestures or nonverbal signals communicate indirectly and acceptably. The group leader should intervene if nonverbal communications are misinterpreted.
- Cultural transference of traits from one person of a certain culture to another person of that culture. If a member has had experiences (usually negative) with people of the same ethnicity as the group leader, the group member may transfer to the leader feelings and reactions developed with others of the leader's ethnicity. To dispel such feelings and reactions, the group leader should detect these misconceptions and reveal them for what they are.
- Cultural countertransference, the group leader's (often subconscious) emotional reaction to a client. Countertransference of culture occurs when a leader's response to a current group member is based on experience with a former group member of the same ethnicity as the new client. Group leaders should exercise restraint in these situations.
- Ethnic prejudice. In multiethnic groups, it is vital to develop an environment in which it is safe to talk about race. Not to do so results in scapegoating or division along racial lines.

In practice, people connect and diverge in ways that cannot be predicted solely on the basis of ethnic or cultural identity. Two people from different ethnic backgrounds may share many other common experiences that provide a basis for identification and mutual support. Leaders are responsible for considering carefully the positions of people who are different in some way, especially when planning fixed-membership groups.

Preparing the Group for New Members
- Inform members in advance that people from a variety of racial and ethnic backgrounds will be in the group
- Discuss the differences at appropriate times in a sensitive way to provide an atmosphere of openness and tolerance
- Set the tone for an open discussion of differences in beliefs and feelings

PP 3-11

To promote group cohesion and welcome new members, the group leader should:

- Inform members in advance that people from a variety of racial and ethnic backgrounds will be in the group.
- Discuss the differences at appropriate times in a sensitive way to provide an atmosphere of openness and tolerance.
- Set the tone for an open discussion of differences in beliefs and feelings.

Preparing the Group for New Members (cont.)
- Help clients adapt to and cope with prejudice in effective ways, while maintaining their self-esteem.
- Integrate new clients into the group slowly, letting them set their own pace.
- When new members start to make comments about others or to accept feedback, encourage more participation.

PP 3-12

- Help clients adapt to and cope with prejudice in effective ways, while maintaining their self-esteem.
- Integrate new clients into the group slowly, letting them set their own pace.
- When new members start to make comments about others or to accept feedback, encourage more participation.

Although arguments for matching the ethnicity of a group leader with that of the group members being treated may have some merit, the reality is that such a course seldom is feasible. Healthcare providers from culturally and linguistically diverse groups are underrepresented in the current service delivery system, so the group leader will likely be from the mainstream culture. Although it might be ideal to match all participants by ethnicity in a therapeutic group, the most important determinants for success are the values and attitudes shared by the group leader and group members.

Other Considerations for Practice
- Expectations of leaders
- Experience in decisionmaking and conflict resolution
- Understanding of gender roles, families, and community
- Values

PP 3-13

Groups may include people who have varying:

- Expectations of leaders (Some cultures might consider leaders problemsolvers, whereas in other cultures leaders might be considered equals until proven otherwise.)
- Experience in decisionmaking and conflict resolution
- Understanding of gender roles, families, and community
- Values

5 minutes

Summary

The trainer:

- *Responds to participants' questions or comments.*
- *Encourages participants to review Chapter 3 of TIP 41.*
- *Instructs participants to read Chapter 4 and reminds them to bring TIP 41 to the next training session.*
- *Reminds participants of the date and time of the next training session.*

Module 4: Group Development and Phase□ Specific Tasks

Module 4 Overview

The goal of Module 4 is to provide participants with an overview of the uses of fixed and revolving groups and an overview of the tasks for the three phases of group development. The information in Module 4 covers Chapter 4 of Treatment Improvement Protocol (TIP) 41, Substance Abuse Treatment: Group Therapy. *This module takes 1 hour to complete and is divided into five sections:*

- *Module 4 Goal and Objectives (5 minutes)*
- *Presentation: Fixed and Revolving Membership Groups (10 minutes)*
- *Presentation: Preparing for Client Participation in Groups (25 minutes)*
- *Presentation: Phase-Specific Group Tasks (15 minutes)*
- *Summary (5 minutes)*

5 minutes

Module 4 Goal and Objectives

After participants have taken their seats, the trainer instructs them to turn to Chapter 4 (p. 59) of TIP 41.

Module 4: Group Development and Phase-Specific Tasks

Based on material in Chapter 4 of TIP 41, *Substance Abuse Treatment: Group Therapy*

PP 4-1

Module 4 covers Chapter 4 of TIP 41.

Module 4 Goal and Objectives

Goal:

Provide an overview of fixed and revolving membership groups and an overview of the tasks for the three phases of group development

Objectives:

- Distinguish the differences between fixed and revolving membership groups
- Prepare clients for groups
- Describe the tasks for each of the three phases of group development

PP 4-2

The goal of Module 4 is to provide an overview of fixed and revolving membership groups and an overview of the tasks for the three phases of group development. By the end of the session, you will be able to:

- Distinguish the differences between fixed and revolving membership groups.
- Prepare clients for groups.
- Describe the tasks for each of the three phases of group development.

10 minutes

Presentation: Fixed and Revolving Membership Groups

Members of fixed membership groups generally stay together for a long time. Members in revolving membership groups remain in the group only until they accomplish their goals. Each is used for different purposes, and each requires different leadership.

Fixed Membership Groups

- Members are prepared and stay together for a long time
- Membership is stable
- Groups are either:
  - Time limited. Members participate in a specified number of sessions and start and finish together
  - Ongoing. New members fill vacancies in a group that continues over a long period
- Fixed groups are rarely used in substance abuse treatment

PP 4-3

Fixed membership groups are small (not more than 15 members) and membership is stable. The group leader usually screens prospective members, who are prepared for participation. Fixed membership groups can be:

- Time limited. The same group of people attends a specified number of sessions, generally starting and finishing together. Learning builds on what has taken place in prior meetings. Members need to be in the group from its start. New members are admitted only in the earliest stages of group development. Time-limited groups are used for skill-building, psychoeducational, and relapse prevention groups.

- Ongoing. New members fill vacancies in a group that continues over a long period. The size of the group is set. The leader generally is less active than is the leader of a time-limited group because interaction among group members is more important than leader-to-member interactions. Leaders need substantial training in group dynamics. Ongoing groups are used for interpersonal process and some psychoeducational groups.

Fixed groups are rare in substance abuse treatment because they demand a long-term commitment of resources.

Revolving Membership Groups

- New members enter a group when they become ready for its services
- Groups must adjust to frequent, unpredictable changes
- Groups are either:
  - Time limited. Member attends a specified number of sessions, starting and finishing at his or her own pace
  - Ongoing. Member remains until he or she has accomplished his or her specified goals
- Revolving groups are frequently used in inpatient treatment programs

PP 4-4

New members enter a revolving membership group when they become ready for the services it provides. Revolving groups must adjust to frequent unpredictable changes. Revolving groups are also:

- Time limited. Each member attends a specified number of sessions, generally starting and finishing at his or her own pace.

- Ongoing. The member remains until he or she has accomplished his or her specified goals.

These groups are frequently found in inpatient treatment programs. Revolving member groups tend to be larger than fixed membership groups. However, if they are larger than 20 members, group interactions break down.

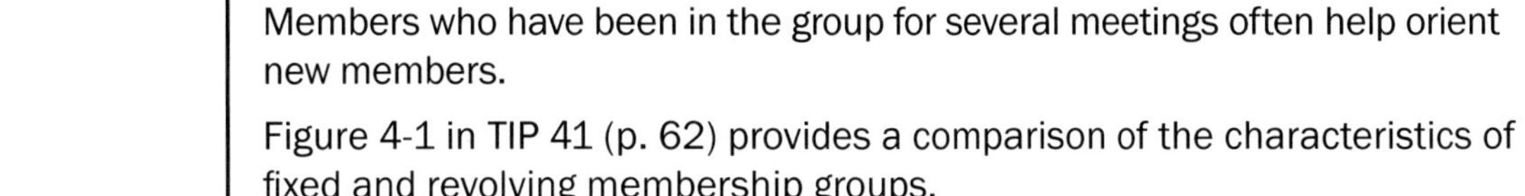

Revolving membership groups are structured and require active leadership. Participation and learning are not highly dependent on previous sessions. Members who have been in the group for several meetings often help orient new members.

Figure 4-1 in TIP 41 (p. 62) provides a comparison of the characteristics of fixed and revolving membership groups.

25 minutes

Pregroup Interviews

- Begin as early as the initial contact between the client and the program
- Strive to:
  - Form a therapeutic alliance between the leader and the client
  - Reach consensus on what is to be accomplished in therapy
  - Educate the client about group therapy
  - Allay anxiety related to joining a group
  - Explain the group agreement

PP 4-5

Preparation Meetings

- Explain how group interactions compare with those in self-help groups
- Emphasize that treatment is a long-term process
- Let new members know they may be tempted to leave the group at times
- Give prospective members an opportunity to express anxiety about group work
- Recognize and address clients' therapeutic hopes

PP 4-6

Presentation: Preparing for Client Participation in Groups

The process of preparing the client for participation in group therapy begins as early as the initial contact between the client and the program. Group leaders should conduct an initial individual session with the candidate for group to form a therapeutic alliance, to reach consensus on what is to be accomplished in therapy, to educate the client about group therapy, to allay anxiety related to joining a group, and to explain the group agreement.

The longer the expected duration of the group, the longer the preparation phase. During this time, the group leader learns how the client handles interpersonal functions, how the client's family functions, and how the client's culture perceives the substance use problem.

Preparation meetings ensure that clients understand expectations and will be able to meet them, and they help clients become familiar with the group therapy process. Client preparation should:

- Explain how group interactions compare with those in self-help groups. Clients should be informed that member-to-member cross-talk, which is discouraged in 12-Step groups, is essential in interactive group therapy.
- Emphasize that treatment is a long-term process. Clients should know that each person's attendance at each session is vital during this process.
- Let new members know they may be tempted to leave the group at times. Clients gain a great deal from persistent commitment to the process and should resist temptations to leave the group. Clients should be encouraged to discuss thoughts about leaving the group as they arise.
- Give prospective members an opportunity to express anxiety about group work. Misperceptions should be countered to keep them from interfering with group participation.
- Recognize and address clients' therapeutic hopes. Leaders can use this information to place clients in groups most likely to fulfill their aspirations.

Leaders should be sensitive to people who are different from the majority of other participants in some way. Clients should always be allowed to be the experts on their own situations.

Leaders are responsible for raising the level of the group members' sensitivity and empathy. They must sometimes prepare group members for a situation in which others have symptoms that could offend or repel them.

Techniques To Increase Retention
- Role induction
- Vicarious pretraining
- Experiential pretraining
- Motivational interviews
- Prompts

PP 4-7

Retention rates are affected positively by client preparation, maximum client involvement during the early stage of treatment, the use of feedback, prompts to encourage attendance, and the provision of wraparound services to make it possible to attend sessions regularly. Consideration needs to be given to the timing and length of groups. To achieve maximum involvement in group therapy, motivational techniques may be used to engage clients.

Techniques that increase retention include the following:

- Role induction uses interviews, lectures, and films to educate clients about the reasons for therapy, setting realistic goals for therapy, expected client behaviors, and so on.
- Vicarious pretraining uses interviews, lectures, films, and other settings to demonstrate what takes place during therapy so that the client can experience the process vicariously.
- Experiential pretraining uses group exercises to teach client behaviors such as self-disclosure and examination of emotions.
- Motivational interviewing uses specific listening and questioning strategies to help clients overcome doubts about making changes.
- Prompts remind clients that upcoming group sessions are important in engaging members during the first 3 months of treatment.

Group Agreements
- Establish the expectations that group members have for one another, the leader, and the group
- Require that group members entering a long-term fixed membership group commit to the group.
- Inspire clients to accept the basic rules and increase their determination and ability to succeed

PP 4-8

A group agreement establishes the expectations that group members have for one another, the leader, and the group. Many leaders require that group members entering a long-term fixed membership group commit to remaining in the group for a set period. The group agreement is intended to inspire clients to accept the basic rules and premises of the group and to increase their determination and ability to succeed.

Elements in a Group Agreement
- Communicating grounds for exclusion
- Confidentiality
- Physical contact
- Use of mood-altering substances
- Contact outside the group
- Participation in the group
- Financial responsibility
- Termination

PP 4-9

A group agreement should include at least eight elements.

- Communicating grounds for exclusion. The terms under which clients will be excluded from the group should be made explicit in the group agreement, so exclusion does not come as a surprise. Some stipulations in the group agreement might have to incorporate legal requirements because court-mandated treatment groups may have attendance criteria set by the State.
- Confidentiality. Group members should be asked not to discuss anything outside the group that could reveal the identity of other members. The leader should emphasize that confidentiality is critical and should encourage members to honor their pledge of confidentiality. The principle that "what's said in the group stays in the group" is a way of delineating group boundaries and increasing trust.

- Physical contact. Touch in a group is never neutral. People have different personal histories and cultural backgrounds that lead to different interpretations of what touch means. Consequently, the leader should evaluate carefully any circumstance in which physical contact occurs, even when it is intended to be positive. In most groups, touch as part of a group ritual is not recommended. Group agreements always should include a clause prohibiting violence.

- Use of mood-altering substances. Some programs, such as those connected to criminal justice systems, have policies that require expulsion of members who are using drugs of abuse. Counselors are required to report these violations. Part of client preparation and orientation is to explain all legally mandated provisions and consequences for failure to comply with treatment guidelines. Members also should pledge to discuss a return to use promptly after it occurs, providing the group rules permit and encourage such disclosures.

- Contact outside the group. Clients who have bonded in group are likely to communicate outside the group and may encounter one another at self-help meetings. Group members need to be told and reminded that new intimate relationships are hazardous to early recovery. Any contacts outside the group should be discussed openly in the group.

- Participation in the group. The agreement should specify what group members are expected to divulge. Group members should be willing to discuss the issues that brought them to group. They should not be required to share personal information until they feel safe enough to do so.

- Financial responsibility. The agreement may specify a commitment to discuss any problems that occur in making payments and the circumstance under which a group member is held responsible for payment. For example, members should know ahead of time that they will be financially responsible for missed sessions if that is the agency policy.

- Termination. Group agreements should specify how group members handle termination. Because group members often are tempted to leave the group prematurely, the agreement should emphasize the need to involve the group in termination decisions. However, members make their own choices about discontinuing treatment.

15 minutes

Beginning Phase:
Preparing the Group To Begin
- Introductions
- Group agreement review
- Providing a safe, cohesive environment
- Establishment of norms
- Initiation of group work

PP 4-10

Presentation: Phase-Specific Group Tasks

Every group has a beginning, middle, and end. These phases occur at different times for different types of groups.

During the beginning phase, the purpose of the group is articulated, working conditions of the group are established, members are introduced, a positive tone is set, and group work begins. This phase may last from 10 minutes to a number of months. In a revolving group, this orientation will happen each time a new member joins the group. Five activities are conducted in the beginning phase:

- Introductions. Even in short-term revolving membership groups, it is important for the leader to connect with each member. All members should give their names and say something about themselves. The leader can build bridges between the old and new members by encouraging old members to help new members join.
- Group agreement review. The group members should review and discuss the group agreement. The leader should ask members whether they have concerns that might require additional provisions to make the group safe. The agreement should be reviewed periodically during the course of the group.
- Providing a safe, cohesive environment. All members should feel that they have a part to play in the group and have something in common. This cohesion affects the productivity of work throughout the therapeutic process.
- Establishment of norms. The group leader is responsible for ensuring that healthful norms are established and that counterproductive norms are precluded, ignored, or extinguished. The leader shapes norms through responses to events in the group and by modeling the behavior expected of others.
- Initiation of group work. The leader facilitates group work by providing information or encouraging honest exchanges among members. Most leaders strive to keep the group focused on the here and now.

Middle Phase:
Working Toward Productive Change
- Both process and content are important
- The group is the forum where clients interact with others
- Clients receive feedback that helps them rethink their behaviors and move toward productive changes
- Leaders allocate time to address issues, pay attention to relations among group members, and model healthful interactions that combine honesty with compassion.

PP 4-11

The group in the middle phase encounters and accomplishes most of the therapy work. During this phase, the leader balances content, which is the information and feelings overtly expressed in the group, and process, which is how members interact in the group. The therapy is both the content and the process. Both contribute to the connections between and among group members.

The group is the forum where clients interact with others. Through give and take, clients receive feedback that helps them rethink their behaviors and move toward productive changes. Leaders allocate time to address issues that arise, pay attention to relations among group members, and model a healthful interactional style that combines honesty with compassion.

End Phase:
Reaching Closure
- Putting closure on the experience
- Examining the impact of the group on each person
- Acknowledging the feelings triggered by departure
- Giving and receiving feedback about the group experience and each member's role in it
- Completing any unfinished business
- Exploring ways to continue learning about topics discussed in the group

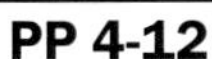

Termination is an important opportunity for members to honor the work they have done, to grieve the loss of associations and friendships, and to look forward to a positive future. Group members learn and practice saying good-bye, understanding that it is necessary to make room in their lives for the next hello.

This phase begins when the group reaches its agreed-on termination point or a member determines it is time to leave the group. Termination is a time for:

- Putting closure on the experience
- Examining the impact of the group on each person
- Acknowledging the feelings triggered by departure
- Giving and receiving feedback about the group experience and each member's role in it
- Completing any unfinished business
- Exploring ways to continue learning about topics discussed in the group

Completing a group successfully can be an important event for group members when they see the conclusion of a difficult but successful endeavor. The longer members have been with the group, the longer they may need to spend on termination.

5 minutes

Summary

The trainer:

- *Responds to participants' questions or comments.*
- *Encourages participants to review Chapter 4 of TIP 41.*
- *Instructs participants to read Chapter 5 and reminds them to bring TIP 41 to the next training session.*
- *Reminds participants of the date and time of the next training session.*

Module 5: Stages of Treatment

Module 5 Overview

The goal of Module 5 is to provide participants with an overview of adjustments that can be made in the early, middle, and late stages of treatment. The information in Module 5 covers Chapter 5 of Treatment Improvement Protocol (TIP) 41, Substance Abuse Treatment: Group Therapy. *This module takes 1 hour to complete and is divided into four sections:*

- *Module 5 Goal and Objectives (5 minutes)*
- *Presentation: Adjustments To Make Treatment Appropriate (5 minutes)*
- *Presentation: The Stages of Treatment (45 minutes)*
- *Summary (5 minutes)*

5 minutes

Module 5 Goal and Objectives

After participants have taken their seats, the trainer instructs them to turn to Chapter 5 (p. 79) of TIP 41.

Module 5: Stages of Treatment

Based on material in Chapter 5 of TIP 41, *Substance Abuse Treatment: Group Therapy*

PP 5-1

Module 5 covers Chapter 5 of TIP 41.

Module 5 Goal and Objectives

Goal:

Provide an overview of adjustments that can be made in the early, middle, and late stages

Objectives:

- Discuss the importance of making adjustments
- Explain the three stages of treatment
- Describe the conditions of the early, middle, and late stages of treatment
- Identify leadership characteristics in the early, middle, and late stages of treatment

PP 5-2

The goal of Module 5 is to provide an overview of adjustments that can made in the early, middle, and late stages of treatment. By the end of the session, you will be able to:

- Discuss the importance of making clinical adjustments in the group therapy.
- Explain the three stages of treatment.
- Describe the conditions of the early, middle, and late stages of treatment.
- Identify leadership characteristics in the early, middle, and late stages of treatment.

5 minutes

Three Stages of Treatment

- In the early stage of treatment, strategies focus on immediate concerns.
- In the middle stage of treatment, clients recognize that their substance abuse causes many problems and blocks them from getting the things they want
- In the last stage of treatment, clients identify the treatment gains to be maintained and risks that remain

PP 5-3

Presentation: Adjustments To Make Treatment Appropriate

Treatment has three stages:

- In the early stage of treatment, clients can be emotionally fragile, ambivalent about relinquishing chemicals, and resistant to treatment. Treatment strategies focus on immediate concerns: achieving abstinence, preventing relapse, and managing cravings. In this stage leaders emphasize hope, group cohesion, and universality.
- In the middle stage of treatment, clients need the group's assistance in recognizing that their substance abuse causes many of their problems and blocks them from getting the things they want. As clients sever their ties with substances, they need help managing their loss and finding healthful substitutes.
- In the last stage of treatment, clients spend less time on their substance abuse issues and turn toward identifying the treatment gains to be maintained and risks that remain. Clients focus on the issues of living, resolving guilt, reducing shame, and adopting a more introspective view of themselves.

Therapeutic strategies change as clients move through the different stages. Interventions that worked well early in treatment may be ineffective and even harmful later in treatment. Stages of recovery and stages of treatment will not correspond perfectly for all people. Clients move in and out of recovery stages in a nonlinear process.

Adjustments in treatment are needed because progress through the stages is not time bound. There is no way to calculate how long an individual will require to resolve the issues that arise in any stage of recovery.

Therapeutic interventions may not account for all of the changes in a particular individual. Generalizations about stages of treatment may not apply to every client in every group.

45 minutes

Condition of Clients in Early Stage of Treatment

- Some enter treatment because of health problems.
- Others begin treatment because they are referred or mandated by the criminal justice system or others.
- Group members are often in extreme emotional turmoil

PP 5-4

Therapeutic Factors in Early Stage of Treatment

- Instilling hope
- Universality
- Imparting information
- Altruism
- Corrective recapitulation of the primary family group
- Developing socializing techniques
- Imitative behavior
- Interpersonal learning
- Group cohesiveness
- Catharsis
- Existential factors

PP 5-5

Presentation: The Stages of Treatment

Typically, people who abuse substances do not enter treatment on their own. Some enter treatment because of health problems; others begin treatment because they are referred or mandated by the criminal justice system, employers, or family members. Group members commonly are in extreme emotional turmoil, grappling with intense emotions such as guilt, shame, depression, and anger about entering treatment. Consequently, the group leader faces the challenge of treating resistant clients. Emphasis is placed on acculturating clients into a new culture, the culture of recovery.

Eleven therapeutic factors contribute to healing as group therapy unfolds:

1. Instilling hope. Some group members exemplify progress toward recovery and support others in their efforts.
2. Universality. Groups enable clients to see that they are not alone and that others have similar problems.
3. Imparting information. Leaders shed light on the nature of addiction via direct instruction.
4. Altruism. Group members gain greater self-esteem by helping one another.
5. Corrective recapitulation of the primary family group. Groups provide a family-like context in which longstanding unresolved conflicts can be revisited and constructively resolved.
6. Developing socializing techniques. Groups give feedback; others' impressions reveal how a client's ineffective social habits might undermine relationships.
7. Imitative behavior. Groups permit clients to try out behaviors of others.
8. Interpersonal learning. Groups correct the distorted perceptions of others.
9. Group cohesiveness. Groups provide a safe environment within which people feel free to be honest and open with one another.
10. Catharsis. Groups liberate clients as they learn how to express feelings and reveal what is bothering them.
11. Existential factors. Groups aid clients in coming to terms with hard truths such as life can be unfair; life can be painful, and death is inevitable; no matter how close one is to others, life is faced alone; it is important to live honestly and not get caught up in trivial matters; each of us is responsible for the ways in which we live.

In different stages of treatment, some of these therapeutic factors receive more attention than others.

Leaders in
Early Stage of Treatment
- Stress that clients have some things in common
- Are spontaneous and engaging
- Are not overly charismatic
- Focus on helping clients:
  - Achieve abstinence
  - Prevent relapse
  - Learn ways to manage cravings

PP 5-6

Clients usually come to the first session of group in an anxious, apprehensive state of mind, which is intensified by the knowledge that they will be revealing personal information and secrets about themselves. The leader begins by stressing that clients have some things in common.

During early treatment, a leader actively engages clients in the treatment process. Clients early on usually respond to group leaders who are spontaneous and engaging. The leader should not be overly charismatic but should be a strong enough presence to meet clients' dependence needs during the early stage of treatment. During early treatment, effective leaders focus on immediate primary concerns:

- Achieving abstinence
- Preventing relapse
- Learning ways to manage cravings

Condition of Clients in
Middle Stage of Treatment
- Clients experience some stability
- Self-knowledge and altruism can be emphasized
- Emotions of anger, sadness, terror, and grief may be expressed more appropriately
- Clients use the group to explore their emotional and interpersonal world

PP 5-7

Cognitive capacity usually begins to return to normal in the middle stage of treatment. A person addicted to cocaine, for example, is dramatically different after 4 months of nonuse. Still, the mind can play tricks. Clients may remember distinctly the comfort of their past substance use yet forget how bad the rest of their lives were and the seriousness of the consequences that loomed before they came into treatment. The temptation to relapse remains a concern.

In the middle stage of treatment, as the client experiences some stability, the therapeutic factors of self-knowledge and altruism can be emphasized. Universality, identification, cohesion, and hope remain important.

As the recovering client's mental, physical, and emotional capacities grow stronger, emotions of anger, sadness, terror, and grief may be expressed more appropriately.

Clients need to use the group as a means of exploring their emotional and interpersonal world. They learn to differentiate, identify, name, tolerate, and communicate feelings.

Therapeutic Strategies in
Middle Stage of Treatment
- Cognitive-behavioral interventions provide tools to modulate feelings and to express and explore them
- Interpersonal groups are helpful

PP 5-8

Cognitive-behavioral interventions can provide clients with specific tools to help modulate feelings and to become more confident in expressing and exploring them. Interpersonal process groups are particularly helpful in the middle stage of treatment because the authentic relationships within the group enable clients to experience and integrate a wide range of emotions in a safe environment.

Leaders in
Middle Stage of Treatment

- Help members see how continued drug or alcohol use interferes with what they want out of life.
- Help clients join the culture of recovery.
- Support the process of change by drawing attention to positive developments.
- Assess the degree of structure and connection clients need as recovery progresses.

PP 5-9

When pointing out contradictions in clients' statements and interpretations of reality, leaders should ensure that confrontations are well timed, specific, and indisputably true. Another way of understanding confrontation is to see it as an outcome rather than as a style. From this point of view, the leader helps group members see how their continued use of drugs or alcohol interferes with what they want to get out of life. In the middle stage, the leader helps clients join a culture of recovery in which they grow and learn. The leader's task is to engage members actively in the treatment and recovery process. To prevent relapse, clients need to learn to monitor their thoughts and feelings, paying attention to internal cues.

The leader can support the process of change by drawing attention to new and positive developments and affirming the possibility of increased connections and new sources of satisfaction. The leader helps individuals assess the degree of structure and connection they need as recovery progresses.

Condition of Clients in
Late Stage of Treatment

- Clients work to sustain the achievements of previous stages.
- Clients may discover and acknowledge that some goals are unrealistic, certain strategies are ineffective, and environments deemed safe are not conducive to recovery.
- Significant underlying issues emerge (e.g., poor self-image, relationship problems, shame, past trauma).

PP 5-10

In the late (also referred to as ongoing or maintenance) stage of treatment, clients work to sustain the achievements of previous stages, but also learn to anticipate and avoid tempting situations and triggers that set off renewed substance use. To deter relapse, the lifestyle that once promoted drinking and drug use are sought out and severed.

Many clients, even those who have reached the late stage of treatment, do return to substance use and an earlier stage of change. Clients who return to substance abuse do so with new information. With it, they may be able to discover and acknowledge that some of the goals they set are unrealistic, certain strategies are ineffective, and environments deemed safe are not at all conducive to successful recovery. With greater insight into the dynamics of their substance abuse, clients are better equipped to make another attempt at recovery and ultimately to succeed.

As the substance use problem fades into the background, significant underlying issues often emerge, such as poor self-image, relationship problems, the experience of shame, or past trauma.

When the internalized pain of the past is resolved, the client will begin to understand and experience healthful mutuality, resolving conflicts without the maladaptive influence of alcohol or drugs. If the underlying conflicts are left unresolved, however, clients are at increased risk of other compulsive behavior, such as excessive exercise, overeating, overspending, gambling, and excessive sexual activity, among others.

Therapeutic Strategies in
Late Stage of Treatment
- The focus of group interaction broadens.
- A process-oriented group may become appropriate for some clients who can confront painful realities.
- The group can be used to settle difficult and painful old business.

PP 5-11

In the early and middle stages of treatment, clients necessarily are so focused on maintaining abstinence that they have little or no capacity to notice or solve other kinds of problems. In late-stage treatment, however, the focus of group interaction broadens. It attends less to the symptoms of drug and alcohol abuse and more to the psychology of relational interactions.

Clients begin to learn to engage in life. As they begin to manage their emotional states and cognitive processes more effectively, they can face situations that involve conflict or emotion. A process-oriented group may become appropriate for some clients who can confront painful realities, such as being abused as a child or being an abusive parent. Other clients may need groups to help them build healthier relationships, communicate more effectively, or become better parents. Some may want to develop job skills. As group members become increasingly stable, they can begin to probe deeper into the relational past. The group can be used to settle difficult and painful old business.

Leaders in
Late Stage of Treatment
- The leader shifts toward interventions that call on clients to take a clear-headed look at their inner world and system of defenses.
- Late-stage interventions permit more intense exchanges.
- The leader allows clients to experience enough anxiety and frustration to identify destructive and maladaptive patterns.

PP 5-12

The leader plays a very different role in the late stage of treatment, which refocuses on helping group members expose and eliminate personal deficits that endanger recovery. Gradually, the leader shifts toward interventions that call on clients to take a clear-headed look at their inner world and system of defenses, which have prevented them from accurately perceiving their self-defeating behavioral patterns. To become adequately resistant to substance abuse, clients should learn to cope with conflict without using substances to escape reality, self-soothe, or regulate emotions.

As in the early and middle stages, the leader helps group members sustain abstinence and makes sure the group provides enough support and gratification to prevent acting out and premature termination. Late-stage interventions permit more intense exchanges. Thus, in late treatment, clients are no longer cautioned against feeling too much. The leader no longer urges them to apply slogans like Turn it over and One day at a time. Clients should manage the conflicts that dominate their lives, predispose them to maladaptive behaviors, and endanger their hard-won abstinence. The leader allows clients to experience enough anxiety and frustration to identify destructive and maladaptive patterns.

5 minutes

Summary

The trainer:

- *Responds to participants’ questions or comments.*
- *Encourages participants to review Chapter 5 of TIP 41.*
- *Instructs participants to read Chapter 6 and reminds them to bring TIP 41 to the next training session.*
- *Reminds participants of the date and time of the next training session.*

Module 6: Group Leadership, Concepts, and Techniques

Module 6 Overview

The goal of Module 6 is to provide participants with an overview of desirable leader traits and behaviors and an overview of the concepts and techniques vital to process groups. The information in Module 6 covers Chapter 6 of Treatment Improvement Protocol (TIP) 41, Substance Abuse Treatment: Group Therapy. *This module takes 1 hour to complete and is divided into four sections:*

- *Module 6 Goal and Objectives (5 minutes)*
- *Presentation: The Group Leader (20 minutes)*
- *Presentation: Concepts and Techniques (30 minutes)*
- *Summary (5 minutes)*

5 minutes

Module 6 Goal and Objectives

After participants have taken their seats, the trainer instructs them to turn to Chapter 6 (p. 91) of TIP 41.

Module 6 covers Chapter 6 of TIP 41.

Module 6: Group Leadership, Concepts, and Techniques

Based on material in Chapter 6 of TIP 41, *Substance Abuse Treatment: Group Therapy*

PP 6-1

Module 6 Goal and Objectives

Goal:

Provide an overview of desirable leader traits and behaviors and an overview of the concepts and techniques vital to process groups.

Objectives:

- Discuss the characteristics of group leaders
- Describe concepts and techniques for conducting substance abuse treatment group therapy

PP 6-2

The goal of Module 6 is to provide an overview of desirable leader traits and behaviors and an overview of the concepts and techniques vital to process groups. By the end of the session, you will be able to:

- Discuss the characteristics of group leaders.
- Describe concepts and techniques for conducting substance abuse treatment group therapy.

20 minutes

Presentation: The Group Leader

When working with people who have substance use disorders, an effective leader uses the same skills, qualities, styles, and approaches needed in any kind of therapeutic group. The particular personal and cultural characteristics of the clients in the group also influence the leader's tailoring of therapeutic strategies to fit the particular needs of the group.

Leaders Choose

- How much leadership to exercise
- How to structure the group
- When to intervene
- How to effect a successful intervention
- How to manage the group's collective anxiety
- How to resolve other issues

PP 6-3

Clients typically respond to a warm, empathic, and life-affirming manner. Group leaders should communicate and share the joy of being alive. This life-affirming attitude carries the unspoken message that a full and vibrant life is possible without alcohol or drugs. The leader is responsible for making a series of choices as the group progresses. The leader chooses:

- How much leadership to exercise
- How to structure the group
- When to intervene
- How to effect a successful intervention
- How to manage the group's collective anxiety
- The means of resolving numerous other issues

It is essential for the leader to be aware of the choices made and to remember that all choices concerning his or her leadership and the group's structure have consequences.

Personal Qualities of Leaders

- Constancy
- Active listening
- Firm identity
- Confidence
- Spontaneity
- Integrity
- Trust
- Humor
- Empathy
  - Communicates respect and acceptance
  - Encourages
  - Is supportive and knowledgeable
  - Compliments
  - Tells less, listens more
  - Gently persuades
  - Provides support

PP 6-4

TIP 41 identifies nine personal qualities of leaders:

- Constancy. An environment with small, infrequent changes is helpful to clients living in the emotionally turbulent world of recovery. Group leaders can emphasize the reality of constancy and security through their behavior. For example, group leaders should always sit in the same place in group, maintain consistent start and end times and ground rules for speaking, and even dress in a consistent manner.
- Active listening. Excellent listening skills are the keystone of any effective therapy. Therapeutic interventions require the leader to perceive and to understand both verbal and nonverbal cues to meaning and metaphorical levels of meaning.

- Firm identity. A firm sense of their own identities, together with clear reflection on experiences in group, enables leaders to understand and manage their own emotional lives. For example, leaders who are aware of their own capacities and tendencies can recognize their own defenses as they come into play in the group. Group work can be intensely emotional. Leaders who are not in control of their own reactions can do significant harm particularly if they cannot admit to a mistake or apologize for it.
- Confidence. Effective leaders operate between the certain and the uncertain. In that zone, they cannot rely on formulas or supply easy answers to clients' complex problems. Instead, leaders have to model the consistency that comes from self-knowledge and clarity of intent, while remaining attentive to each client's experience and the unpredictable unfolding of each session's work. This secure grounding enables the leader to model stability for the group.
- Spontaneity. Good leaders are creative and flexible.
- Integrity. Leaders should be familiar with their institution's policies and with pertinent laws and regulations. Leaders also need to be anchored by clear internalized standards of conduct and able to maintain the ethical parameters of their profession.
- Trust. Leaders should be able to trust others.
- Humor. Leaders need to be able to use humor appropriately, which means that it is used only in support of therapeutic goals and never is used to disguise hostility or wound anyone.
- Empathy. One of the cornerstones of successful group therapy for substance abuse, empathy is the ability to identify someone else's feelings while remaining aware that the feelings of others are distinct from one's own. For the counselor, the ability to project empathy is an essential skill. An empathic leader:
  - ñ Communicates respect for and acceptance of clients and their feelings.
  - ñ Encourages in a nonjudgmental, collaborative relationship.
  - ñ Is supportive and knowledgeable.
  - ñ Sincerely compliments rather than denigrates or diminishes another person.
  - ñ Tells less and listens more.
  - ñ Gently persuades but understands that the decision to change is the client's.
  - ñ Provides support throughout the recovery process.

One of the feelings that the leader needs to empathize with is shame, which is common among people with substance use histories. Shame is so powerful that it should be addressed whenever it becomes an issue.

Group therapy with clients who have histories of substance abuse requires active, responsive leaders who keep the group lively and on task and ensure that members are engaged continuously and meaningfully with one another.

Leading Groups
- Leaders vary therapeutic styles to meet the needs of clients
- Leaders model behavior.
- Leaders can be cotherapists.
- Leaders are sensitive to ethical issues:
  - Overriding group agreement
  - Informing clients of options
  - Preventing enmeshment
  - Acting in each client's best interest

PP 6-5

- Leaders vary therapeutic styles to meet the needs of clients. During the early and middle stages of treatment, the leader is active, becoming less so in the late stage. To determine the type of leadership required to support a client in treatment, the leader should consider the client's capacity to manage affect, level of functions, social supports, and stability. These considerations are essential to determine the type of group best suited to meet the client's needs.

- Leaders model behavior. It is more useful for the leader to model group-appropriate behaviors than to assume the role of mentor. Leaders should be aware that self-disclosure is always occurring, whether consciously or subconsciously. They should use self-disclosure only to meet the task-related needs of the group and only after thoughtful consideration. When personal questions are asked, leaders need to consider the motivation behind the question. Often clients are seeking assurance that the leader understands and can assist them.

- Leaders can be cotherapists. Cotherapy (also called coleadership) is an effective way to blend diverse skills, resources, and therapeutic perspectives that two therapists can bring to the group. It provides an opportunity to watch adaptive behavior. A maleñfemale cotherapy team may be especially helpful; it shows people of opposite sexes engaging in a healthful, nonexploitative relationship.

- Leaders are sensitive to ethical issues. As the group process unfolds, the leader needs to be alert, always ready to perceive and resolve issues with ethical dimensions. Typical situations with ethical concerns include:

  ñ Overriding group agreement. Group agreements give the group definition and clarity and are essential for group safety. In rare situations, however, it would be unethical not to bend the rules to meet the needs of an individual. Sometimes the needs of the group override courtesies shown to an individual. For example, if a member becomes seriously ill and must miss sessions, other members may want to express their concerns for the missing member in the group even though they have agreed not to discuss absent members.

  ñ Informing clients of options. Even when group participation is mandated, clients should be informed of options open to them. For example, a member deserves the option to discuss with an administrator a leader's behavior that the client finds inappropriate.

  ñ Preventing enmeshment. Leaders should be aware that the power of groups can have a dark side. The need to belong is so strong that it can sometimes cause a client to act in a way that is not genuine or consistent with personal ethics. The leader needs to monitor group sharing to ensure that clients are not drawn into situations that violate their privacy or integrity; the leader is obligated to respect the rights and best interests of individuals.

– Acting in each client's best interest. It is possible that the group collectively may validate a particular course of action that may not be in a client's best interest. The leader is responsible for challenging conclusions or recommendations that deny individual autonomy or could lead to serious negative consequences.

Leading Groups (cont.)
- Leaders handle emotional contagion
  - Protect individuals
  - Protect boundaries
  - Regulate affect
- Leaders work within professional limitations
- Leaders ensure flexibility in clients' roles
- Leaders avoid role conflict

PP 6-6

- Leaders handle emotional contagion. Another's sharing can stir frightening memories and intense emotions in listeners. In this atmosphere, leaders are required to:

  ñ Protect individuals. The leader should guard the right of each member to refrain from involvement, making it clear that each member has the right to private emotions and feelings.

  ñ Protect boundaries. No one should be obligated to share intimate details.

  ñ Regulate affect. The leader needs to modulate affect (emotionality), always keeping it at a level that enables the work of the group to continue.

- Leaders work within professional limitations. Leaders should never attempt to use techniques for which they are not trained or with populations or in situations for which they are unprepared. When new techniques are used or new populations are being served, leaders should have appropriate training and supervision.

- Leaders ensure flexibility in clients' roles. Although it is natural for members to assume certain roles (one client may naturally take on the role of a leader, whereas another may assume the role of scapegoat), individual members benefit from experiencing aspects of themselves in different roles. Role variation keeps the group lively and dynamic.

- Leaders avoid role conflict. Leaders should be sensitive to issues of dual relationships. A leader's responsibilities outside the group that place him or her in a different relationship to group participants should not be allowed to compromise the leader's in-group role. For example, leaders should avoid attending self-help meetings at which group members are present.

Leading Groups (cont.)
- Leaders improve motivations when:
  - Members are engaged at the appropriate stage of change
  - Members receive support for change efforts
  - The leader explores choices and their consequences with members
  - The leader communicates care and concern for members
  - The leader points out members' competencies
  - Positive changes are noted in and encouraged by the group

PP 6-7

- Leaders improve motivations. Motivation generally improves when:

  ñ Members are engaged at the appropriate stage of change.

  ñ Members receive support for change efforts.

  ñ The leader explores choices and their consequences with members.

  ñ The leader communicates care and concern for members.

  – The leader points out members' competencies.

  ñ Positive changes are noted in and encouraged by the group.

Leading Groups (cont.)
- Leaders work with, not against, resistance
- Leaders protect against boundary violations
- Leaders maintain a safe, therapeutic setting
  - Emotional aspects of safety
  - Substance use
  - Boundaries and physical contact
- Leaders help cool down affect
- Leaders encourage communication within the group

PP 6-8

- Leaders work with, not against, resistance. Resistance generally arises as a defense against the pain that therapy and examining one's behavior usually brings. In group therapy, resistance is at the individual and group levels. The leader should have a repertoire of means to overcome resistance that prevents success.

- Leaders protect against boundary violations. Providing a safe, therapeutic frame for clients and maintaining firm boundaries are among the most important functions of the leader. The boundaries established should be mutually agreed on and specified in a group agreement or the ground rules. Boundary violations should be pointed out in a nonjudgmental, matter-of-fact way.

- Leaders maintain a safe, therapeutic setting:

  - Emotional aspects of safety. Members should learn to interact in positive ways. They need to feel safe without blaming or scapegoating others. If a member makes an openly hostile comment, the leader should clearly state that members are not to be attacked.

  - Substance use. The presence of an intoxicated group member will upset many members. The leader should ask the person who has relapsed to leave the current session. After the person has left, group members can explore feelings about the relapse and affirm their abstinence.

  - Boundaries and physical contact. When physical boundaries are breached in the group and no one in the group raises the issue, the leader should remind the members of the terms of the agreement and call attention to the questionable behavior in a straightforward way. A leader should know the agency's policies regarding violent behavior. Members should be allowed to opt out of activities that involve physical contact.

- Leaders help cool down affect. Leaders carefully monitor the level of emotional intensity in the group, recognizing that too much too fast can bring on extremely uncomfortable feelings that interfere with progress. When intervening to control runaway affect, leaders should be careful to support the genuine expressions of emotions that are appropriate for the group and the individual's stage of change.

- Leaders encourage communication within the group. Leaders' primary task is stimulating communication among group members, rather than between individual members and the leader. Leaders can do this by praising good communication, noticing a person's body language and asking the person to express the feeling, and helping members know that their contributions are important. Leaders should speak often but briefly.

30 minutes

Presentation: Concepts and Techniques

Interventions may be directed to an individual or the group. They can be used to clarify what is going on, redirect energy, stop a process that is not helpful, or help the group decide what should be done. A well-timed, appropriate intervention has the power to:

Interventions
- Connect with other people.
- Discover connections between substance use and thoughts and feelings.
- Understand attempts to regulate feelings and relationships.
- Build coping skills.
- Perceive the effect of substance use on life.
- Notice inconsistencies among thoughts, feelings, and behavior.
- Perceive discrepancies.

PP 6-9

- Help a client connect with other people.
- Discover connections between the use of substances and inner thoughts and feelings.
- Understand attempts to regulate feelings and relationships.
- Build coping skills.
- Perceive the effect of substance use on one's life.
- Notice meaningful inconsistencies among thoughts, feelings, and behavior.
- Perceive discrepancies between stated goals and what is actually being done.

Avoiding a Leader-Centered Group
- Build skills in members; avoid doing for the group what it can do for itself.
- Encourage group members to learn the skills necessary to support and encourage one another.
- Refrain from overresponsibility for clients. Clients should be allowed to struggle with what is facing them.

PP 6-10

To move away from the center stage, leaders can:

- Build skills in members; avoid doing for the group what it can do for itself.
- Encourage the group to learn the skills necessary to support and encourage one another. Too much or too frequent support from the leader can lead to approval seeking, which blocks growth and independence. Supporting one another is a skill that should develop through group phases.
- Refrain from overresponsibility for clients. Clients should be allowed to struggle with what is facing them.

Confrontation
- Can have an adverse effect on the therapeutic alliance and process.
- Can point out inconsistencies such as disconnects between behaviors and stated goals.
- Can help clients see and accept reality, so they can change accordingly.

PP 6-11

Group leaders have come to recognize that, when confrontation is equivalent to attack, it can have an adverse effect on the therapeutic alliance and process, ultimately leading to failure. A more useful way to think about confrontation is pointing out inconsistencies such as disconnects between behaviors and stated goals. Confrontation used this way is part of the change process and part of the helping process. It helps clients see and accept reality, so they can change accordingly.

Transference and Countertransference
- Transference: Clients project parts of important past relationships into present relationships.
- Countertransference: The leader projects emotional response to a group member's transference:
  - Feelings of having been there
  - Feelings of helplessness when the leader is more invested in the treatment than the clients are
  - Feelings of incompetence because of unfamiliarity with culture and jargon

PP 6-12

Transference means that people project parts of important past relationships into present relationships.

Emotions inherent in groups are not limited to clients. The groups inevitably stir up strong feelings in leaders. Countertransference is the group therapy leader's emotional response to a group member's transference. Examples include:

- Feelings of having been there. Leaders with histories of substance abuse have an extraordinary ability to empathize with clients who abuse substances. If that empathy is not understood or controlled, it can become a problem if the group leader tries to act as a role model or discloses too much personal information.
- Feelings of helplessness when the leader is more invested in the treatment than the clients are.
- Feelings of incompetence because of unfamiliarity with culture and jargon. The leader should ask clients to explain terms and expressions that can be misunderstood.

The leader needs to manage all feelings associated with countertransference. With the help of supervision, the leader can use countertransference to support the group process.

Resistance
- Resistance arises to protect the client from the pain of self-examination and change
- Effective leaders welcome resistance as an opportunity to understand something important for the client or the group
- Leaders may have contributed to the resistance.
- Efforts need to be made to understand the problem

PP 6-13

Resistance arises as a client's subconscious defense to protect himself or herself from the pain of self-examination and change. These processes within the client or group impede the open expression of thoughts and feelings or block the progress of an individual or group. The effective leader will neither ignore resistance nor attempt to override it. Instead, the leader helps the individual or group understand what is getting in the way, welcoming the resistance as an opportunity to understand something important for the client or the group.

Leaders should recognize that members are not always aware of their reasons for resistance. They should explore what is happening and what can be learned from it, not battle the resistance. Leaders may have contributed to the resistance, and efforts need to be made to understand the problem.

Confidentiality
- Strict adherence to confidentiality regulations builds trust
- Leaders should explain how information from sources may and may not be used in group.
- Violations of confidentiality should be managed in the same way as other boundary violations

PP 6-14

Strict adherence to confidentiality regulations builds trust. If the bounds of confidentiality are broken, serious legal, personal, and professional consequences may result. All group leaders should be thoroughly familiar with Federal laws on confidentiality and relevant agency policies. Leaders should warn clients that what they say in group may not be kept strictly confidential.

Group leaders have many sources of information on a client, including the client's employers and spouse. Leaders should clearly explain how information from these sources may and may not be used in group.

Violations of confidentiality among members should be managed in the same way as other boundary violations; that is, empathic joining with those involved followed by a factual reiteration of the agreement that has been broken and an invitation to group members to discuss their perceptions and feelings.

Integrating Care

- Integrations with other healthcare professionals. Professionals in the healthcare network need to be aware of the role of group therapy.
- Integrations of group therapy and other forms of therapy. Clinicians should coordinate the treatment plan, keeping important interpersonal issues alive in both settings.
- Medication knowledge base. Leaders should be aware of medication needs of clients, the types of medications prescribed, and side effects.

PP 6-15

Professionals within the entire healthcare network need to become more aware of the role of group therapy for people abusing substances. To build the understanding needed to support people in recovery, group leaders should educate others serving this population as often as opportunities arise, such as when clinicians from different sectors of the healthcare system work together on a case. Similar needs for understanding exist with probation officers, families, and primary care physicians.

It is common for a client to be in both individual and group therapy simultaneously. The dual relationship creates both problems and opportunities. Skilled counselors can use what they discover in group about the client's style of relatedness to enhance individual therapy. In situations in which one counselor sees a client individually and another treats the same client in group, the counselors should be in close communication with each other. They should coordinate the treatment plan, keeping important interpersonal issues alive in both settings. The client should know that this collaboration routinely occurs for the client's benefit.

Leaders should be aware of various medication needs of clients, the types of medications prescribed, and potential side effects. The pregroup interview for long-term groups should ask each group member what medications he or she is taking and the names of prescribing physicians so cooperative treatment is possible. (Consent forms for the sharing of information should be signed when appropriate.) If an evaluation of prescription medications is needed, counselors should refer the client to a consulting physician working with the agency or to a physician knowledgeable about chemical dependence. Attention needs to be paid to medications prescribed for physical illnesses as well.

Handling Conflict
- Conflict is normal, healthful, and unavoidable.
- Handling anger, developing empathy, managing emotions, and disagreeing respectfully are major tasks.
- The leader facilitates interactions between members in conflict and calls attention to subtle, unhealthful patterns.
- Conflicts that appear to scapegoat a group member may be misplaced anger that a member feels toward the leader.

PP 6-16

Conflict in group therapy is normal, healthful, and unavoidable. It can present opportunities for group members to find meaningful connections with one another and within their lives.

Handling anger, developing empathy for a different viewpoint, managing emotions, and working through disagreements respectfully are major and worthwhile tasks for recovering clients. The leader's judgment and management are crucial as these tasks are handled. It is just as unhelpful to clients to let the conflict go too far as it is to shut down a conflict before it gets worked through. The group leader must gauge the verbal and nonverbal reactions of every group member to ensure that everyone can manage the emotional level of the conflict.

The leader facilitates interactions between members in conflict and calls attention to subtle, unhealthful patterns. Conflicts within groups may be overt or covert. The leader helps the group bring covert conflicts into the open. The observation that a conflict exists and that the group needs to pay attention to it actually makes group members feel safer. The decision to explore the conflict further is made based on whether such inquiry would be productive.

Leaders should be aware that many conflicts that appear to scapegoat a group member are actually misplaced anger that a member feels toward the group leader. When the leader suspects this situation, the possibility should be forthrightly presented to the group.

Individual responses to particular conflicts can be complex and can resonate powerfully according to a client's values. After a conflict, it is important for the group leader to speak privately with members and determine how each feels. Leaders often use the last 5 minutes of a session in which a conflict has occurred to give group members an opportunity to express their concerns.

Subgroup Management
- Subgroups inevitably will form.
- Subgroups can provoke anxiety, especially when a therapy group comprises individuals acquainted before becoming group members.
- Subgroups are not always negative.

PP 6-17

In any group, subgroups inevitably will form. Individuals will feel more affinity and more potential alliance with some members than they feel with others. One key role for the leader is to make covert alliances overt.

Subgroups can sometimes provoke anxiety, especially when a therapy group comprises individuals who were acquainted before becoming group members. Such connections are potentially disruptive. When groups are formed, leaders should consider whether subgroups would exist.

When subgroups stymie full participation, the group leader can reframe what the subgroup is doing. At other times a change in room arrangement may be able to reconfigure undesirable combinations.

Subgroups are not always negative. The leader may intentionally foster a subgroup that helps marginally connected clients move into the life of the group.

Responding to Disruptive Behavior
- Clients who cannot stop talking
- Clients who interrupt
- Clients who flee a session

PP 6-18

TIP 41 identifies three types of disruptive behaviors:

- Clients who cannot stop talking. When a client talks on and on, he or she may not know what is expected in a group. At other times, a client may talk more than his or her share because he or she is not sure of what else to do. If group members exhibit no interest in stopping a compulsive talker, it may be appropriate to examine this silent cooperation. Group members may be avoiding examining their own past patterns of substance abuse and forging a more productive future. When this motive is suspected, the leader should explore what group members have and have not done to signal the speaker that it is time to yield the floor. It may also be advisable to help the talker find a more effective strategy for being heard.
- Clients who interrupt. Interruptions disrupt the flow of discussion in the group, with frustrating results. The client who interrupts is often someone new to the group and not yet accustomed to its norms and rhythms.
- Clients who flee a session. Clients who run out of a session often are acting on an impulse that others share. It would be productive in such instances to discuss these feelings with the group and to determine what members can do to talk about these feelings. The leader should stress that the therapeutic work requires members to remain in the room and talk about problems instead of attempt to escape them.

Contraindications for Continued Participation
- Sometimes, clients are unable to participate in ways consistent with group agreements.
- Removing someone from group is serious and should never be done without careful thought and consultation.
- The leader makes the decision to remove an individual from the group.
- Members are allotted time to work through their responses.

PP 6-19

Sometimes, clients are unable to participate in ways consistent with group agreements. They may attend irregularly, come to group intoxicated, show little or no impulse control, or fail to take medications to control a co occurring disorder. Removing someone from group is serious and should never be done without careful thought and consultation.

The decision to remove an individual is not one the group makes. The leader makes the decision and explains to the group why the action was taken. Members then are allotted time to work through their responses to what is bound to be a highly charged event. Anger at the leader for acting without group input or acting too slowly is common in expulsion situations and should be explored.

Managing Common Problems
- Coming late or missing sessions
- Silence
- Tuning out
- Participating only around the issues of others
- Fear of losing control
- Fragile clients with psychological emergencies
- Anxiety and resistance after self-disclosure

PP 6-20

TIP 41 briefly addresses seven common problems in group:

- Coming late or missing sessions. Sometimes leaders view the client who comes to group late as a person who is behaving badly. It is more productive to see this kind of boundary violation as a message to be deciphered.

- Silence. Nonresponsiveness may provide clues to clients' difficulties with connecting with their inner lives or with others. Special consideration is sometimes necessary for clients who speak English as a second language (ESL). Experiences involving strong feelings can be hard to translate. When feelings are running high, even fluent ESL speakers may not be able to find the right words to say what they mean or may be unable to understand what another group member is saying about an intense experience.

- Tuning out. When clients seem present in body but not in mind, it helps to tune into them just as they are tuning out. The leader should explore what was happening as an individual became inattentive.

- Participating only around the issues of others. Even when group members are disclosing little about themselves, they may be gaining a great deal from the group experience, remaining engaged on issues that others bring up. To encourage a member to share more, a leader might introduce the topic of how well members know one another and how well they want to be known.

- Fear of losing control. Sometimes clients avoid opening up because they may become tearful in front of others. When this restraint becomes a barrier, the leader should help them remember ways that they have handled strong feelings in the past. When a client cries or breaks down, the leader should validate the feelings and should concentrate on the person's adaptive abilities.

- Fragile clients with psychological emergencies. Because clients know that the leader is bound to end the group's work on time, they often wait until the last few minutes to share emotionally charged information. The leader should recognize that the client has deliberately chosen this time to share this information. The timing could be the client's way of limiting the group's response and avoiding an onslaught of interest. The group should point out this self-defeating behavior and encourage the client to change it.

- Anxiety and resistance after self-disclosure. Clients may feel great anxiety after disclosing something important. Leaders should assure clients that people disclose information when they are ready and that they do not have to reiterate the disclosure when new clients enter the group.

5 minutes

Summary

The trainer:

- *Responds to participants' questions or comments.*
- *Encourages participants to review Chapter 6 of TIP 41.*
- *Instructs participants to read Chapter 7 and reminds them to bring TIP 41 to the next training session.*
- *Reminds participants of the date and time of the next training session.*

Module 7: Supervision

Module 7 Overview

The goal of Module 7 is to provide participants with an overview of the skills group therapy leaders need, the purpose and value of clinical supervision, and necessary training. The information in Module 7 covers Chapter 7 of Treatment Improvement Protocol (TIP) 41, Substance Abuse Treatment: Group Therapy. *This module takes 45 minutes to complete and is divided into four sections:*

- *Module 7 Goal and Objectives (5 minutes)*
- *Presentation: Training (15 minutes)*
- *Presentation: Supervision (10 minutes)*
- *Summary (15 minutes)*

5 minutes

Module 7 Goal and Objectives

After participants have taken their seats, the trainer instructs them to turn to Chapter 7 (p. 123) of TIP 41.

Module 7: Supervison

Based on material in Chapter 7 of TIP 41, *Substance Abuse Treatment: Group Therapy*

PP 7-1

Module 7 covers Chapter 7 of TIP 41.

Module 7 Goal and Objectives

Goal: Provide an overview of the skills group therapy leaders need, the purpose and value of clinical supervision, and necessary training

Objectives:

- Identify training opportunities.
- Appreciate the value of clinical supervision.

PP 7-2

The goal of Module 7 is to provide an overview of the skills group therapy leaders need, the purpose and value of clinical supervision, and necessary training. By the end of the session, you will be able to:

- Identify training opportunities.
- Appreciate the value of clinical supervision.

15 minutes

Presentation: Training

Common Errors

- Impatience with clients' slow pace of dealing with changes
- Inability to drop the mask of professionalism
- Failure to recognize countertransference issues
- Not clarifying group rules
- Conducting individual therapy rather than using the entire group effectively
- Failure to integrate new members effectively into the group

PP 7-3

Many substance abuse treatment counselors have not had specific training and supervision in the special skills needed to be an effective group leader. Common errors that counselors make include:

- Impatience with clients' slow pace of dealing with changes
- Inability to drop the mask of professionalism
- Failure to recognize countertransference issues
- Not clarifying group rules
- Conducting individual therapy rather than using the entire group effectively
- Failure to integrate new members effectively into the group

Training and education for group leaders working in the substance abuse treatment field can alleviate or eliminate such errors. Simultaneously, additional training is becoming even more critical because the traditionally separate fields of mental health and substance abuse counseling increasingly overlap, requiring more and more cross-knowledge, and because an ever-younger pool of clients is presenting with more cognitive deficits, abuse issues, and co-occurring disorders.

Staff Development Needs

- Theories and techniques
  - Traditional psychodynamic methods, cognitive–behavioral modes, systems theory
- Observation
  - Sitting in on groups, studying tapes, watching groups
- Experiential learning
  - Participating in a training group, being a member of a personal therapy group, or joining in experiences
- Supervision
  - Ongoing training with groups under the supervision of an experienced leader

PP 7-4

A group leader for people in substance abuse treatment requires competencies in both group work and addiction. Before leading a group, a leader needs training in:

- Theories and techniques. Theories may include traditional psychodynamic methods, cognitiveñbehavioral modes, and systems theory. Applications that pertain to a wide variety of settings and particular client populations are drawn from these theoretical bases.
- Observation. The observer can sit in on group therapy sessions, study videotapes of group sessions (ordinarily followed by a discussion), or watch groups live through one-way mirrors.
- Experiential learning. The leader can participate in a training group offered by an agency, become a member of a personal therapy group (these are often process-oriented groups), or join in group experiences at conferences.

- Supervision. Training is ongoing with groups under the supervision of an experienced leader. Supervision in a group enables leaders to obtain first-hand experiences and helps them better understand what is happening in groups that they will eventually lead.

In group therapy with clients with substance use disorders, establishing and maintaining credibility with all group members can be challenging. Leaders not in recovery will need to anticipate and respond to group members' questions about their experience with substances and will need skills to handle group dynamics focused on this issue. However, leaders in recovery may tend to focus too much on themselves. Group leaders emotionally invested in acting as models of recovering perfection are setting themselves up for negative group dynamics.

The main issue is not whether the leader is in recovery. What matters most is whether the leader is trained in group therapy and addiction treatment and has good judgment and leadership skills. Helping the group explore why the recovery status of the group leader is important can be discussed when members raise the issue.

Training Opportunities
- American Group Psychotherapy Association (AGPA) http://www.agpa.org
- American Psychiatric Association http://www.psych.org
- American Psychological Association http://www.apa.org
- American Society of Addiction Medicine (ASAM) http://www.asam.org
- Association for Specialists in Group Work (ASGW) http://asgw.org

PP 7-5

National professional organizations are a rich source of training. Through conferences or regional sessions, national associations provide both experiential and didactic training geared to the needs of a wide range of professionals, from the novice to the highly experienced counselor. Professional organizations that provide a variety of training include:

- American Group Psychotherapy Association (AGPA) has more than 4,000 members who provide professional, educational, and social support for group counselors: http://www.agpa.org.

- American Psychiatric Association has more than 35,000 U.S. and international member physicians who work to ensure humane and effective treatment for all persons with mental disorders: http://www.psych.org.

- American Psychological Association offers a certificate of proficiency in the treatment of alcohol and other psychoactive substance use disorders: http://www.apa.org.

- American Society of Addiction Medicine (ASAM) educates health professionals about addiction: http://www.asam.org.

- Association for Specialists in Group Work (ASGW) is a division of the American Counseling Association and was founded to promote high-quality training, practice, and research: http://asgw.org.

Training Opportunities (cont.)

- Association for the Advancement of Social Work with Groups (AASWG) http://www.aaswg.org
- NAADAC—National Association for Addiction Professionals http://www.naadac.org
- National Association of Black Social Workers (NABSW) http://www.nabsw.org
- National Association of Social Workers (NASW) http://socialworkers.org

PP 7-6

Association for the Advancement of Social Work with Groups (AASWG) is an international organization that develops standards to reflect the distinguishing features of group work: http://www.aaswg.org.

- NAADAC National Association for Addiction Professionals is the largest organization for alcoholism and drug abuse professionals and offers workshops, seminars, and education programs for members: http://www.naadac.org.
- National Association of Black Social Workers (NABSW) offers conferences and mentoring programs to support the work of African-American social workers: http://www.nabsw.org.
- National Association of Social Workers (NASW) has developed practice standards and clinical indicators, a credentialing program, continuing education courses, and publications: http://socialworkers.org.

Federal Resources

- Order TIPs, TAPs, and mental health information from SAMHSA at 1-877-SAMHSA-7 (1-877-726-4727) or at http://www.store.samhsa.gov
- Download publications from CSAT's Knowledge Application Program at http://www.kap.samhsa.gov
- National Institute on Drug Abuse provides materials at http://www.nida.nih.gov

PP 7-7

Many agencies mandate a certain number of trainings each year and provide in-house training that draws on the resources of credentialed senior management. Each State has a department of alcohol and drug abuse services, and some States provide substance abuse training for group therapy. Training in mental health issues is often available through the mental health division of government agencies, professional associations, and psychological and psychiatric organizations. Most colleges, universities, and community colleges offer relevant courses.

Two Federal entities offer resources for training. The Substance Abuse and Mental Health Services Administration (SAMHSA) provides a number of resources, including the TIP Series and Technical Assistance Publications (TAPs). Publications in these series can be ordered from the SAMHSA Store at 1-877- SAMHSA-7 (1-877-726-4727) or at http://www.store.samhsa.gov. Publications can also be downloaded from the Knowledge Application Program Web site at http://www.kap.samhsa.gov. The SAMHSA Store also provides a wealth of information on mental health issues. The National Institute on Drug Abuse provides materials at http://www.nida.nih.gov.

10 minutes

Supervision

- Essential skills include:
  - Substance abuse treatment
  - Group training
  - Cultural competence
  - Diagnosis of co-occurring disorders
- Supervisory alliance is required to teach skills needed to lead groups and ensure that the group accomplishes its purposes.

PP 7-8

Presentation: Supervision

Clinical supervision, as it pertains to group therapy, is best carried out within the context of group supervision. Group dynamics and group process facilitate learning by setting up a microcosm of a larger social environment.

The supervisor should be competent in several content areas, including substance abuse treatment, group training, cultural competence, and diagnosis of co-occurring disorders. A supervisor may be an administrator, an in-house trainer, or a counselor from another agency.

The key to effective group therapy supervision is the development of the supervisory alliance. In this working relationship, the counselor develops skills in group analysis and refines abilities to develop appropriate treatment strategies. The supervisory alliance is needed to teach the counselor the skills needed to lead groups effectively and to make sure that the group accomplishes its purposes.

Assessment of Leader Skills
- Clinical skills (from selecting prospective group members and designing treatment strategies to planning and managing termination)
- Comprehensive knowledge of substance abuse, which could entail broad general knowledge of, or a thorough facility with, a particular field
- Knowledge of the preferred theoretical approach
- Knowledge of psychodynamic theory
- Knowledge of the institution's preferred theoretical approaches

PP 7-9

The supervisor should be able to assess the domains that leaders are required to master. These include:

- Clinical skills (from selecting prospective group members and designing treatment strategies to planning and managing termination)
- Comprehensive knowledge of substance abuse, which could entail broad general knowledge of, or a thorough facility with, a particular field
- Knowledge of the preferred theoretical approach
- Knowledge of psychodynamic theory
- Knowledge of the institution's preferred theoretical approaches

---

Assessment of Leader Skills (cont.)
- Diagnostic skills for determining co-occurring disorders
- Capacity for self-reflection, recognizing one's vulnerability and ability to monitor reactions
- Consultation skills, the ability to consult with referring therapist, provide feedback, and coordinate both individual and group treatment
- Capacity to be supervised, including openness in supervision, setting goals for training, and discussing one's learning style and preferences

PP 7-10

- Diagnostic skills for determining co-occurring disorders
- Capacity for self-reflection, such as recognizing one's own vulnerability and ability to monitor and govern behavioral and emotional reactions
- Consultation skills, such as the ability to consult with the referring therapist, provide feedback, and coordinate both individual and group treatment
- Capacity to be supervised, including openness in supervision, setting goals for training, and discussing with a supervisor one's learning style and preferences

---

15 minutes

Summary

In this last session, the trainer:

- *Responds to participants' questions or comments.*
- *Asks participants for feedback on the course in general and any modules that they found particularly interesting.*
- *If using feedback/course evaluation forms, asks participants to complete forms before they leave.*
- *Thanks participants for attending the training and encourages them to explore further the issues raised during the training.*

Made in United States
North Haven, CT
24 July 2023